Advances in Anatomy Embryology and Cell Biology

Vol. 140

Springer
*Berlin
Heidelberg
New York
Barcelona
Budapest
Hong Kong
London
Milan
Paris
Santa Clara
Singapore
Tokyo*

Andreas Miething

The Establishment of Spermatogenesis in the Seminiferous Epithelium of the Pubertal Golden Hamster

(Mesocricetus auratus)

With 34 Figures and 6 Tables

Springer

Andreas Miething
University of Bonn
Institute of Anatomy
Nussallee 10
53115 Bonn
Germany

ISBN 3-540-63655-2 Springer-Verlag Berlin Heidelberg New York

Library of Congress-Catalog-Card-Number
Miething, Andreas, 1955- . The establishment of spermatogenesis in the seminiferous epithelium of the pubertal golden hamster (Mesocricetus auratus) / p. cm. -- (Advances in anatomy, embryology, and cell biology; Vol. 140) Includes bibliographical references (p.) and index.
ISBN 3-540-63655-2 (soft : alk. paper) 1. Golden hamster—Spermatozoa. 2. Golden hamster—Development. 3. Spermatogenesis in animals. I. Title. II. Series: Advances in anatomy, embryology, and cell biology; V. 140.
QL801.E67 Vol. 140 [QL737.R666] 571 s—dc21 [571.8'45]

Printed in Germany

Production: PRO-EDIT GmbH, D-69126 Heidelberg
SPIN: 10633376 27/3136-5 4 3 2 1 0 - Printed on acid-free paper

For my family
Bernadette, Lukas, Lorenz,
and Jolanthe

Acknowledgements

I wish to express my sincere appreciation to Professor Dr. H. Wartenberg (Bonn) for stimulating discussions and for his interest in the present study. I am also indebted to Professor Dr. S.G. Haider (Düsseldorf) for his valuable advice regarding the enzyme histochemistry of Leydig cells. The excellent technical assistance of Mrs. M. Sonnenwald, Mrs. W. Langmann, and Mr. L. Droese is gratefully acknowledged. Mrs. V. Viebahn kindly reviewed the English manuscript.

Contents

1 Introduction

Within mammalian ontogenesis, the reproductive system reaches full maturation and adult function later than any other organ system. This is particularly true of the gonads and their key role in reproduction. Although significant propagation of germ cells occurs in the immature male gonad, and the population of fetal-type interstitial cells (Leydig cells) exercises an essential influence on the development of the endocrine and phenotype sex, testicular maturation, leading up to male fertility, is not completed until puberty. It includes the successive establishment of the process of spermatogenesis within the emerging seminiferous epithelium of the testicular tubules and, at the same time, the final maturation of Leydig cells. This period is the subject of the present study.

In the following, the testicular development up to the onset of the pubertal period is briefly summarized (Sect. 1.1), thus defining the starting point of the period investigated in this study. In the subsequent section (Sect. 1.2), a brief survey of the organization of adult spermatogenesis is presented, focusing on its spatial and temporal arrangement, its hormonal control, and the significance of the somatic elements within the seminiferous epithelium. Knowledge of this basic data constitutes the background required for the systematic analysis of the developmental phenomena of the pubertal seminiferous epithelium.

1.1 Fetal and Prepubertal Development of the Testis

Gonadal development in mammals, including humans, comprises a complex pattern of growth and differentiation that is divided into several phases. It begins with the formation of the genital ridge and growth of the gonadal anlagen in the dorsal area of the coelomic epithelium. The gonads are still sexually indifferent during this early period, but are already composed of two main cellular components, i.e., the somatic cells and the immigrated primordial germ cells. Sexual differentiation (testis/ovary) is first detectable in the testis, in which the somatic tissue is rearranged into a network of solid, epithelial cords which also enclose the moderately proliferating germ cells. Around the time of sexual differentiation, testicular development enters the period of prespermatogenesis (Hilscher et al. 1974; Wartenberg 1976), which represents a complex interplay of cellular differentiation and proliferation and involves both somatic and germ cell development within the seminiferous cords. In several small mammalian species (e.g., rat, mouse, hamster), the germ cells, now referred to as prospermatogonia according to Hilscher et al. (1974), pass through clearly separated, successive phases of enhanced mitotic proliferation (M- and T_2-prospermatogonia) or quiescence (T_1-

prospermatogonia), respectively (e.g., Hilscher and Makoski 1968; Hilscher et al. 1974; Wartenberg 1976; Pützer 1986; Miething 1993). In these species, M-, T_1-, and T_2-prospermatogonial phases each last only a few days, and the final transformation of T_2-prospermatogonia into the first "mature" stem cells of spermatogenesis, the spermatogonia, marks the transition from prespermatogenesis to the initiation of spermatogenesis. Other mammalian species, such as rabbit, cattle, and especially humans, exhibit a less clear pattern of prospermatogonial development. In these species, in different areas within the seminiferous cords, germ cells may enter periods of proliferation or mitotic quiescence at various times and therefore display heterogeneous states of differentiation.

In contrast, in the female, the germ cells (oocytes) enter meiosis at the corresponding developmental time after only a short phase of rapid oogonial proliferation (Wartenberg 1974). This comparative view highlights the main features of prespermatogenic testicular development, i.e., mitotic propagation of the germ cell population and concomitant impediment of the onset of male germ cell meiosis (Hilscher 1983; Wartenberg 1990).

However, not all authors designate and interpret the prepubertal development of male germ cells according to the above concept of prespermatogenesis defined by Hilscher et al. (1974) and Wartenberg (1976). Rather, the phase of restoring prospermatogonial proliferation after the T_1-period of mitotic quiescence is often regarded as the starting point of adult spermatogenesis and equated with the appearance of adult-type spermatogonia (e.g., Kluin and de Rooij 1981; Vergouwen et al. 1991, 1993). However, this view does not take into account the specific morphological and topographic features of T_2-prospermatogonia, which allow a clear distinction from spermatogonia.

The time scale of the cardinal steps of testicular development in the golden hamster, including the period of prespermatogenesis, is summarized in Fig. 1. In this species, prespermatogenesis lasts from day 10 post conception (dpc 10) beyond the time of birth on dpc 16 up to day 12 post partum (dpp 12). It is on this day that the first "mature" A-spermatogonia appear.

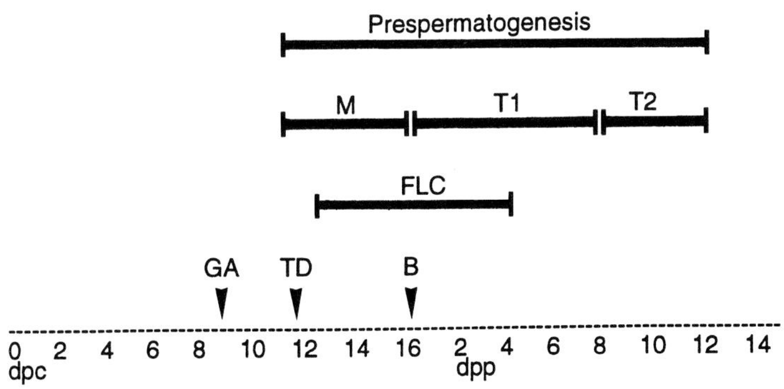

Fig. 1. Survey of testicular development in the golden hamster up to the onset of puberty. *dpc*, days post conception; *dpp*, days post partum; *M*, M-prospermatogonia (Wartenberg 1976; Pützer 1986; Miething 1993); *T1*, T_1-prospermatogonia (Wartenberg 1976; Pützer 1986; Miething 1989, 1993); *T2*, T_2-prospermatogonia (Wartenberg 1976; Miething 1989, 1993); *FLC*, fetal Leydig cells (Gondos et al. 1974); *GA*, gonadal anlagen (Boyer 1953); *TD*, testicular differentiation (Price and Ortiz 1965; Pützer 1986; Miething 1993); *B*, birth (Price and Ortiz 1965; Viswanathan and Davis 1992)

1.2 Adult Spermatogenesis

The process by which a small number of spermatogonia give rise to a great quantity of spermatozoa (Roosen-Runge 1974), i.e., spermatogenesis, can be subdivided into three distinct phases: the proliferation of spermatogonia, the meiotic divisions of spermatocytes, and the morphological and functional maturation of spermatids (Clermont 1972). The successive steps of germ cell development in the golden hamster (Clermont 1954) can be followed in Fig. 2, which depicts the consecutive types of germ cells line by line from the bottom to the top.

Three generations of A-spermatogonia (A_1, A_2, A_3) are followed by In- and then B-spermatogonia. Mitoses of the latter give rise to primary spermatocytes, which, after accomplishing a last (premeiotic) S phase, enter meiosis I. This comprises an exceptionally prolonged prophase, including the successive leptotene, zygotene, pachytene, and (short) diplotene phases and the rapidly completed meta-, ana-, and telophase. The resulting secondary spermatocytes immediately pass through meiosis II to produce spermatids. The morphogenetic process by which young spermatids evolve to spermatozoa (spermiogenesis) is characterized by three simultaneously occurring events: nuclear condensation and shaping, development of the acrosome, and development of the flagellum. Subdivision of this process into 17 steps in the golden hamster (Clermont 1954) is mainly based on the morphology of the emerging acrosome.

Most of the above-mentioned mitotic and all meiotic germ cell divisions display an incomplete cytokinesis. By means of the resulting intercellular bridges, daughter germ cells remain permanently linked to one another and thus run through spermatogenesis as synchronously developing germ cell clones (Fawcett et al. 1959; Fawcett 1961; Gondos 1973, 1984).

Another general feature of mammalian spermatogenesis is a certain amount of germ cell loss by degeneration (Regaud 1900; Roosen-Runge 1955, 1973; Huckins 1972, 1978; Kerr 1992). Affecting the different types of germ cells to varying extents, it is thought to contribute to a considerable reduction in the overall cellular output of spermatogenesis (Clermont 1962; Barr et al. 1971; Kerr 1992). The significance of this phenomenon is, however, still a matter of speculation; it might lie in a negative selection of (chromosomally) aberrant cells (Oakberg 1956a; Roosen-Runge 1973) or might represent a regulative mechanism designed to maintain a constant numerical ratio of germ cells to Sertoli cells (Huckins 1978) and to compensate for a local overproduction of germ cells (de Rooij and Lok 1987), respectively. A significant increase in germ cell degeneration (affecting preleptotene and leptotene spermatocytes as well as certain spermatids) has been observed in gonadotropin-deficient rats (Russell and Clermont 1977; Russell et al. 1981; Ghosh et al. 1991). A comparable hormone-dependent degenerative response has also been reported for the germ cells of the hamster and mouse (Russell et al. 1990). In several recent studies on this subject, attention has been paid in particular to the mode of degeneration of the germ cells involved, which may be necrosis or apoptosis (see Kerr et al. 1972; Wyllie et al. 1980). Apoptotic cell death has been attributed to certain types of germ cells (Allan et al. 1987, 1992; Furuchi et al. 1996; Henriksen et al. 1996) as well as to conditions of gonadotropin or androgen deficiency (Tapanainen et al. 1993; Sinha Hikim et al. 1995; Billig et al. 1995; Brinkworth et al. 1995).

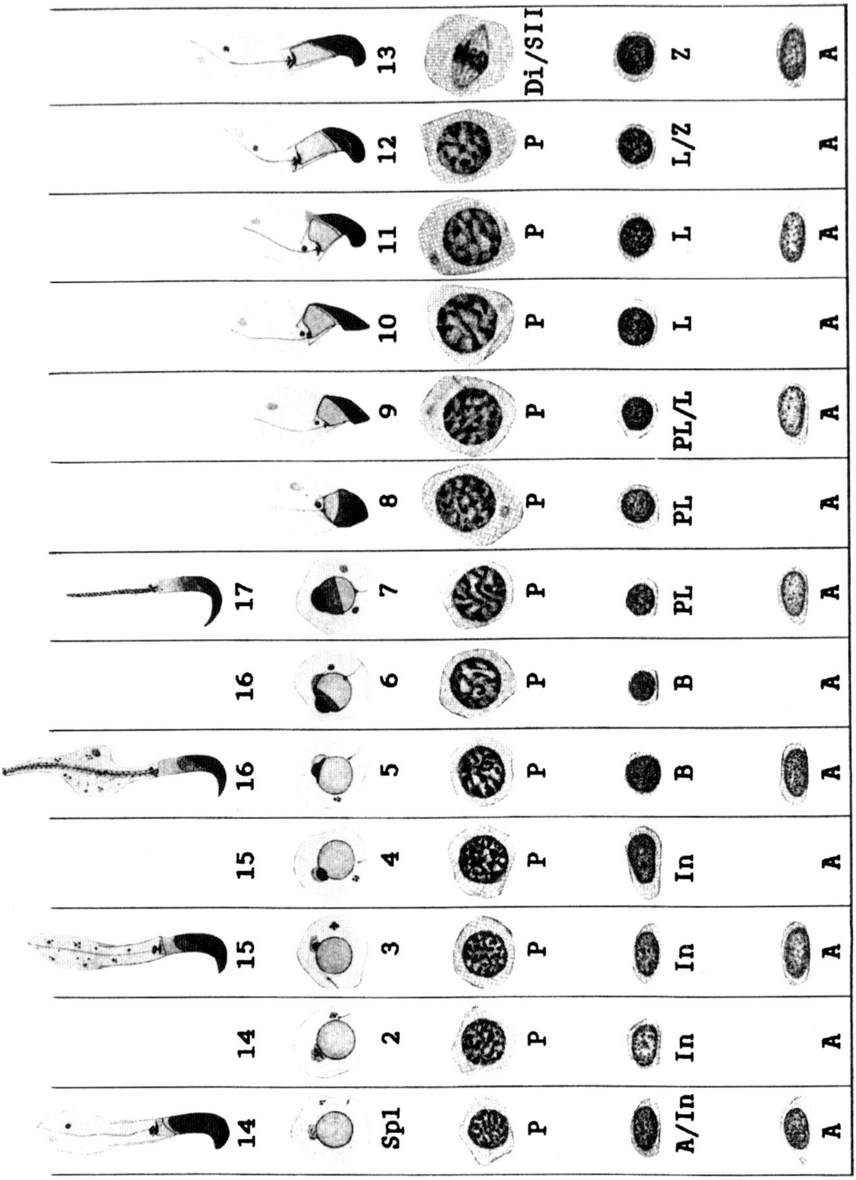

Fig. 2. Cycle map of spermatogenesis in the golden hamster based upon original illustrations by Clermont (1954) used by kind permission. *I–XIII*, stages I–XIII of spermatogenesis; *A*, A-spermatogonia; *In*, intermediate spermatogonia; *B*, B-spermatogonia; *PL*, postmitotic/preleptotene spermatocytes I; *L*, leptotene spermatocytes I; *Z*, zygotene spermatocytes I; *P*, pachytene spermatocytes I; *Di*, diplotene spermatocytes I; *SII*, spermatocytes II; *Sp1–17*, spermatids of steps 1–17 of spermiogenesis

One feature that is particularly interesting for the present study is the spatial organization of the spermatogenic process within the seminiferous epithelium. Most mammalian species studied (including rodents such as the rat, mouse, and golden hamster) display a comparatively clear organizational pattern of germ cell development; within a given local area of the seminiferous epithelium (cross-sectioned tubule), there is defined grouping of the germ cell types present at particular phases of development. In the course of time, such cell associations or *stages of spermatogenesis*, of which there is a limited number, follow each other in a cyclic manner. The periodical reappearance of the complete sequence of the stages constitutes the *cycle of the seminiferous epithelium*.

The stages of spermatogenesis represent artificial subdivisions of the continuous spermatogenic process (Clermont 1972). While the cyclic character of spermatogenesis has long been recognized (von Ebner 1871, 1888; Brown 1885; Benda 1887; Regaud 1901), the stages were not defined in detail until much later on the basis of different criteria. These were either the morphology and position of spermatids (e.g., Roosen-Runge and Giesel 1950; Roosen-Runge 1952; Swierstra and Foote 1963), the characteristics of the changing acrosomal system of spermatids as seen in periodic acid–Schiff (PAS)–hematoxylin-stained paraffin sections (e.g., Leblond and Clermont 1952; Clermont 1954; Oakberg 1956a; Clermont and Perey 1957b), or the proliferative pattern of spermatogonia (Huckins 1978).

Spermatogenesis in the golden hamster has been divided into 13 stages (Clermont 1954), designated by the Roman numerals I–XIII. Originally defined on the basis of PAS–hematoxylin-stained paraffin sections, this classification is also evident in toluidine blue/pyronine-stained semithin sections (Wartenberg and Schubert 1976; Schubert 1977). The resulting cycle map of spermatogenesis is presented in Fig. 2.

In addition to the *vertical organization* of the seminiferous epithelium summarized above, there is also a *horizontal organization*, i.e., the topographical arrangement of the stages along the longitudinal course of the tubules. A regular alternation of secretory periods (of the spermatozoa) in different segments of the tubules was noted as early as 1887 by Benda and has been interpreted as a prerequisite for the continuous production of spermatozoa. On the basis of an analysis of longitudinal sections of paraffin-embedded testicular tubules of adult rats, Perey et al. (1961) gave an elaborate description of the horizontal organization along the tubule length. Briefly, these authors identified the *segments* of the tubules that are occupied by single germ cell associations, and they defined the *wave of the seminiferous epithelium* as the series of segments along the tubule that comprises all the consecutive stages of the seminiferous cycle, including local modulations of the segmental order (Perey et al. 1961). Whether this pattern of organization is definitively established during early development (embryological theory) or whether it appears later and is maintained as a result of interactions between neighboring cells (coordination theory) (Perey et al. 1961) has not yet been clarified.

Spermatogenesis in humans and in a few other primate species is not arranged in the sense of the above-mentioned linear wave of the seminiferous epithelium, but has been found to follow a complex topographical pattern based on a helical arrangement of germ cell populations (Schulze and Rehder 1984; Schulze et al. 1986).

Another aspect of spermatogenic organization is the species-specific, largely invariable duration of the cycle of the seminiferous epithelium and of its single stages. Though not yet entirely understood, the stable temporal pattern of spermatogenesis is probably due to the highly constant duration of the successive steps of germ cell proliferation and differentiation and, in particular, requires newly arising generations of germ cells to enter the spermatogenic process at temporally fixed intervals (Clermont 1972). In the golden hamster, a cycle of the seminiferous epithelium has been reported to last between 8.7 (Clermont and Trott 1969) and 9.0 days (de Rooij 1968). Durations of the single stages range from 6.3 to 37.2 h (de Rooij 1968; Clermont and Trott 1969). The complete process of spermatogenesis takes approximately 35 days (Clermont and Trott 1969).

Besides the germ cells (and closely related to them), the somatic Sertoli cells are also found within the seminiferous epithelium. Initially characterized by Sertoli (1865), both their complex morphology and biochemistry and their various functions with regard to spermatogenesis have been extensively studied since then (for reviews, see Fawcett 1975; Steinberger and Steinberger 1977; Russell 1980; Schulze 1984; Tindall et al. 1985; Jegou 1993; Russell and Griswold 1993). The main functional aspects of Sertoli cells are briefly addressed below:

- By means of their numerous membrane attachments to the basement membrane and to the various germ cells, Sertoli cells maintain the epithelial character of the seminiferous tissue and allow specific cell-to-cell communication (e.g., Connell 1974; Russell 1977a,b, 1993; Russell et al. 1990; Vogl et al. 1993).
- The characteristic basolateral contact zones linking adjacent Sertoli cells constitute the Sertoli cell (or blood–testis) barrier, which permanently subdivides the seminiferous epithelium into a basal and an adluminal compartment (Dym and Fawcett 1970; Fawcett et al. 1970; Gilula et al. 1976; Pelletier and Friend 1983; Pelletier and Byers 1992). In the course of their basoapical ascent, germ cells pass across this barrier immediately before entering meiosis I (preleptotene spermatocytes) (Russell 1977a, 1978; Byers et al. 1993).
- The formation and maintenance of a tubular lumen is assumed to depend on the secretion of fluid by the Sertoli cells in the presence of a functional Sertoli cell barrier (Vitale et al. 1973; Russell et al. 1989; Gondos and Berndtson 1993).
- Sertoli cells participate substantially in the delivery of nutrients to the germ cells, especially to those within the adluminal compartment (Sertoli 1865; Peter 1899; Vilar et al. 1962; Griswold 1988, 1993).
- A variety of Sertoli cell secretion products, such as several growth factors, inhibin, androgen-binding protein, and luteinizing hormone-releasing hormone (LHRH)-like factor, are thought to be involved in regulation of germ cell development throughout spermatogenesis (see Sharpe et al. 1982; Griswold 1988; de Jong 1988; Skinner 1991, 1993).
- Phagocytotic activity of Sertoli cells is highly developed and directed to degenerating germ cells as well as to tubulobulbar complexes and residual bodies during spermiation (Regaud 1901; Dietert 1966; Kerr and de Kretser 1974; Fawcett 1975; Russell and Clermont 1976; Chemes 1986; Pineau et al. 1991).
- Sertoli cells actively participate in the cyclic organization of the seminiferous epithelium. This concerns several structural and cytoskeletal features, the periodical reformation of the Sertoli cell barrier, endocytotic and phagocytotic activities, and secretion of various Sertoli cell products (e.g., Russell 1977a, 1978; Schulze 1984; Amlani and Vogl 1988; Ye et al. 1993; Morales and Clermont 1993; Parvinen 1993; Zhu et al. 1997).

Spermatogenesis is controlled by the two pituitary gonadotropins luteinizing hormone (LH) and follicle-stimulating hormone (FSH). LH acts indirectly by stimulating the Leydig cells within the testicular interstitial tissue to produce and release testosterone, which in turn promotes spermatogenesis. Both Sertoli cells (Tindall et al. 1977) and different types of germ cells (Wright and Frankel 1980; Vornberger et al. 1994) contain receptors for testosterone. The second gonadotropin, FSH, directly stimulates the seminiferous epithelium. FSH receptors are present on the surface of Sertoli cells and probably also on the surface of spermatogonia (Means et al. 1976; Orth and Christensen 1977). Furthermore, FSH is able to enhance the effects of LH and testosterone by increasing the number of LH receptors on the surface of Leydig cells and by stimulating the secretion of androgen-binding protein by Sertoli cells, respectively. However, the actual significance of FSH for the maintenance of adult spermatogenesis is still not fully understood. In addition to hormonal regulation, various local paracrine factors seem to be involved in regulating the complex interactions between the cell populations of the seminiferous epithelium, the peritubular cells, and the Leydig cells (Sharpe 1986; Spiteri-Grech and Nieschlag 1993). The existence of local factors exerting an inhibiting effect on the proliferation of spermatogonia (chalones) remains controversial (Clermont and Mauger 1974; Cunningham and Huckins 1979a; Bustos-Obregon et al. 1988).

1.3 Pubertal Establishment of Spermatogenesis

The developmental interval between termination of prespermatogenesis and qualitatively complete spermatogenesis comprises the period during which the germ cells initially pass through the entire sequence of spermatogenic differentiation from the A-spermatogonium to the mature spermatid. By the time the first spermiation occurs, the seminiferous epithelium is already adult in qualitative terms. Fertility, however, is not reached until several days later, when spermatozoa have passed the epididymis and have attained the full capability to fertilize an egg (Horan and Bedford 1972; Weissenberg et al. 1994). In the course of pubertal development of the testis, maturation of Sertoli cells, peritubular cells, and Leydig cells is also completed. This occurs in close morphological and functional association with germ cell development.

The present study examines in detail the process by which mammalian spermatogenesis is established, using the golden hamster as a model. The organization and progress of this final period of testicular maturation is of interest in itself due to the tightly harmonized development of all cellular and tissue components involved. Furthermore, the analysis of spermatogenesis, which is in the process of being established, might eventually also reveal new aspects of cellular interrelations and regulative mechanisms operative in adult spermatogenesis.

The study takes the morphological progress of germ cell development as a point of reference to analyze the organizational pattern and level of synchronization of the initial spermatogenic cycles. This also includes degenerative phenomena and the proliferative behavior of the germ cells. In addition, several exemplary developmental parameters of Sertoli cells (Sertoli cell barrier, proliferation) and Leydig cells (activity of 11β-hydroxysteroid dehydrogenase, 11β-HSD) are examined. In the discussion, these maturational features of somatic cells and the developmental progression of germ cells proceeding in immediate temporal or spatial proximity will be reviewed and their close relationship emphasized.

2 Material and Methods

2.1 Animals

Testes of male golden (Syrian) hamsters (*Mesocricetus auratus*, HsdHan:AURA) ranging in age from 12 to 38 dpp were studied. The animals were kept under standard laboratory conditions with a cycle of 14 h light and 10 h darkness. The day following the night of birth (usually after a pregnancy of 16 days) was defined as day 0. The developmental period between days 12 and 22 was studied at daily intervals, and the subsequent period up to day 38 at intervals of 2 days. This resulted in an overall number of 19 successive age-groups; two animals were used for each. In addition, testicular tissue from several adult male golden hamsters aged between 2 and 6 months was included in the study.

2.2 Histology

Animals were anesthetized with sodium pentobarbitol; following an injection of heparin (Sprando 1990), they were perfusion-fixed with a mixture of 2% glutaraldehyde and 2% paraformaldehyde in 0.1 *M* phosphate buffer (Karnovsky 1965) via the left ventricle. Testes were removed and pieces of 1–2 mm in length per side were immersed in the above fixative for 2 h and postfixed in 1% phosphate-buffered osmium tetroxide for 4 h. In order to demonstrate the existence of the Sertoli cell barrier, a primary fixative containing a mixture of the above aldehydes in 0.05 *M* cacodylate buffer and an additional 0.5–0.8 *M* lanthanum nitrate was applied for at least 18 h. After dehydration, the specimens were embedded in araldite. Flat embedding of isolated seminiferous tubules was carried out as described previously (Miething 1996). Semithin sections (1 μm) were stained with toluidine blue/pyronine (Ito and Winchester 1963; modified according to Holstein and Wulfhekel 1971). Selected semithin sections were reembedded for subsequent thin sectioning (Schabtach and Parkening 1974; Larramendi 1985). Thin sections (50–90 nm) were contrasted with uranyl acetate and lead citrate (Stempak and Ward 1964; Reynolds 1963) and examined under an electron microscope (Philips EM 301).

2.3
Immunohistochemistry

Immunohistochemical detection of the thymidine analogue bromodeoxyuridine (BrdU) incorporated into S-phase cells (Gratzner 1982; Dolbeare 1995) was employed to evaluate cell kinetics. Two hours before commencement of the embedding procedure, animals received an i.p. injection of BrdU (Sigma, Heidelberg, Germany) dissolved in 0.9% saline at a dose of 250 μg/g body weight. Tissue samples were fixed and embedded as described in Sect. 2.2 (postfixation in osmium tetroxide was omitted). Serial semithin sections were cut alternately at 1.5 μm (for immunohistochemistry) and 1.0 μm (for morphological coanalysis after toluidine blue/pyronine staining). To remove the resin, sections were pretreated in sodium methylate for 5–15 min (Mayor et al. 1961; modified according to Drenckhahn and Franz 1986) followed by hydration through acetone and graded ethanols to phosphate-buffered saline (PBS). For immunohistochemical staining, the monoclonal antibody anti-BrdU (Bu20a; Dako, Hamburg,Germany) and the alkaline phosphatase anti-alkaline phosphatase (APAAP) procedure were used (Cordell et al. 1984). Except for the trypsin treatment, all the incubation steps specified below were carried out at room temperature. The procedure included the following:

- Proteolytic predigestion by 0.1% trypsin (in 0.1% calcium chloride, 20 min at 37°C)
- DNA denaturation in 4 *N* HCl (20 min) and subsequent neutralization with 0.1 *M* borax solution (pH 8.5, 15 min)
- Anti-BrdU antibody diluted 1:30 in PBS (30 min)
- Monoclonal goat anti-mouse IgG (Dako) 1:60 in PBS (30 min)
- APAAP complex (Dako) 1:30 in PBS (30 min)
- Alkaline phosphatase substrate, containing, in 0.1 *M* TRIS buffer, naphthol AS-MX phosphate, fast red TR salt, and levamisole to inhibit endogenous enzyme activity (5–10 min in the dark)
- Rinsing in PBS and distilled water and mounting of sections in Mowiol (Hoechst, Frankfurt, Germany)

In negative control slides, PBS was used instead of anti-BrdU antibody.

Determination of S-phase indices (percentage of BrdU/APAAP-labeled cells) of germ and Sertoli cells was performed as follows: for each developmental state studied (adult state and all developmental steps of the pubertal seminiferous epithelium), labeled cells within 20 tubule cross-sections were counted and labeling indices were evaluated separately for the respective cell types involved by directly comparing the nuclei in the BrdU/APAAP-marked sections and in the corresponding, neighboring toluidine blue/pyronine-stained sections. Statistical characterization included calculation of standard errors of mean (SEM).

2.4
Enzyme Histochemistry

Animals were deeply anesthetized (Sect. 2.2) and testes were removed. Specimens of testicular tissue were transferred to gelatin capsules containing OCT cryomedium (Miles Scientific, Elkhart/Indiana, USA) and immediately frozen in liquid nitrogen.

Cryostat sections (10 μm) were prepared, put on glass slides covered with poly-l-lysine, and processed for enzyme histochemical reaction of 11β-HSD according to the tetrazolium method (Lojda 1965; Lojda et al. 1979; Haider 1988):

- Fixation in acetone (5 min at 4°C)
- Incubation for 90–120 min at 37°C in the dark in a medium containing the substrate 11β-hydroxytestosterone (Sigma, Deisenhofen, Germany), β-Nicotinamide adenine dinucleotide (NAD), tetranitroblue tetrazolium chloride (TNBT; Merck, Darmstadt, Germany), potassium cyanide, polyvinylalcohol (20%), and 0.1 *M* phosphate buffer.

After rinsing in distilled water, sections were counterstained with 0.1% nuclear fast red and mounted in Mowiol (Hoechst, Frankfurt, Germany). Negative controls were incubated in a medium lacking substrate.

3 Results

3.1 Adult Seminiferous Epithelium

Knowledge of the morphology of adult germ cells and of their characteristic cellular associations in the 13 adult spermatogenic stages serves as basis for the systematic analysis of the developing pubertal seminiferous epithelium. The stages and types of germ cells composing them (defined by Clermont 1954) are documented in Figs. 3 (stages I–IV), 4 (stages V–VIII), 5 (stages IX–XII), and 6 (stage XIII). Progress of spermatogenesis and stepwise maturation of germ cells may be followed most easily by comparing the micrographs of the seminiferous epithelium at the different stages (Figs. 3–6) with their respective representations in the cycle map of Fig. 2.

3.2 Developmental Steps of the Pubertal Seminiferous Epithelium

At first sight, histological evaluation of the pubertal testicular tissue on any of the days studied reveals considerable differences regarding the developmental state of the seminiferous epithelium, and in particular of the germ cells, in the various tubule sections. This diversity of dissimilarly developed tubule segments will first be defined, followed by a systematic analysis of the temporally proceeding pubertal maturation of the seminiferous epithelium that superimposes the above-mentioned developmental differences.

3.2.1 Terminology

The terminology for the different developmental steps of the seminiferous epithelium proposed below is based on the developmental level reached by the germ cells present. Irrespective of the age of the testicular tissue, the current state of pubertal spermatogenesis is specified as exactly as possible. As the maturational progress of pubertal germ cells principally follows the adult pattern of germ cell differentiation, the cycle map of adult spermatogenesis (see Fig. 2) is used for orientation. If this map is

Fig. 3a–d. Light micrographs of **a** stage I, **b** stage II, **c** stage III, and **d** stage IV of adult spermatogenesis. ×700. *A*, A-spermatogonium; *In*, In-spermatogonium; *P*, pachytene spermatocytes; *1–4, 14, 15*, spermatids of steps 1–4, 14, and 15

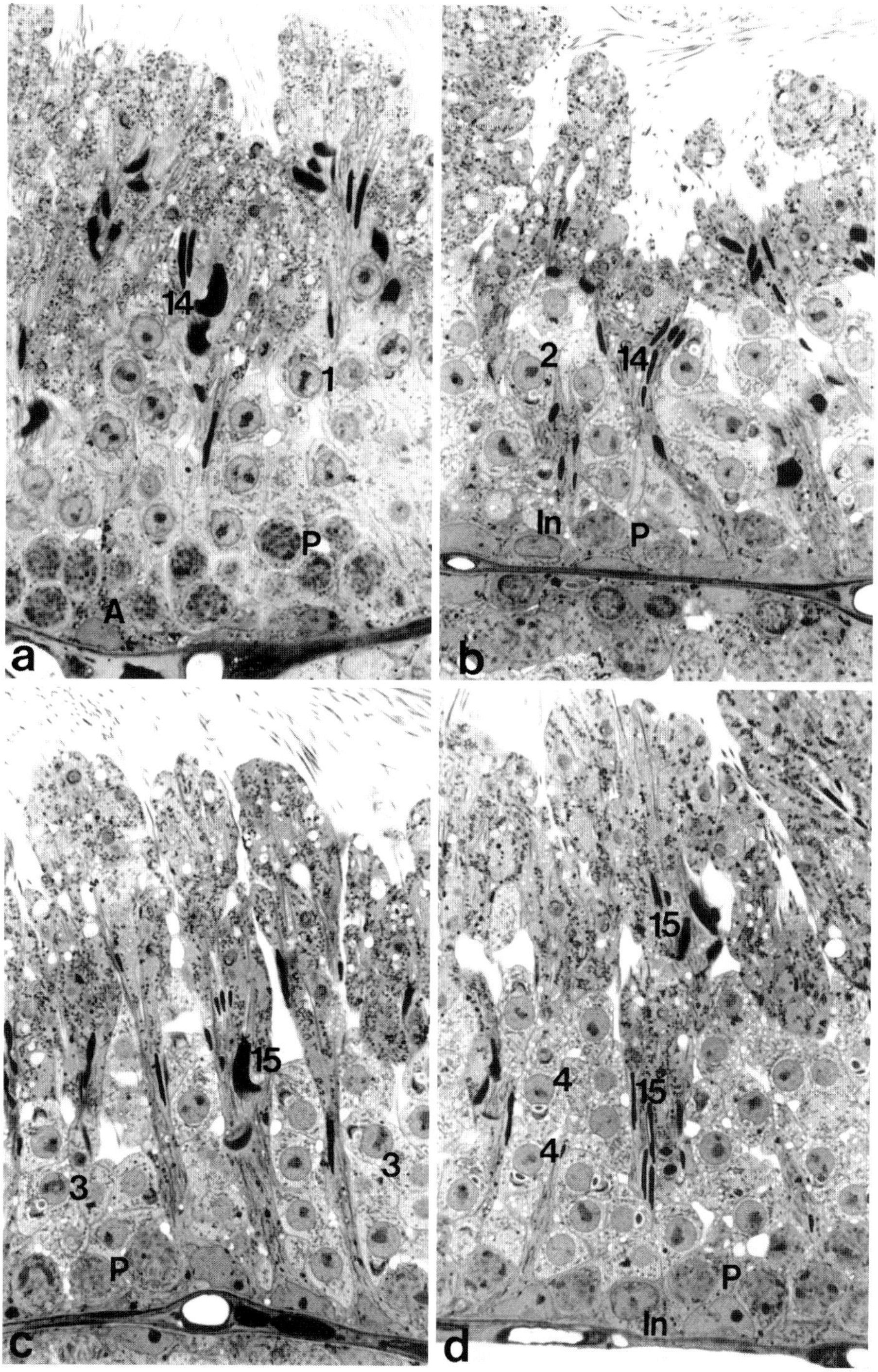
14
1
P
A
a
2
14
In
P
b
15
3
3
P
c
15
15
4
4
P
In
d

Fig. 4a–d. Light micrographs of **a** stage V, **b** stage VI, **c** stage VII, and **d** stage VIII of adult spermatogenesis. ×700. *B*, B-spermatogonium; *PL*, postmitotic/preleptotene spermatocytes; *P*, pachytene spermatocytes; *5–8, 16, 17*, spermatids of steps 5–8 , 16, 17; *R*, residual bodies

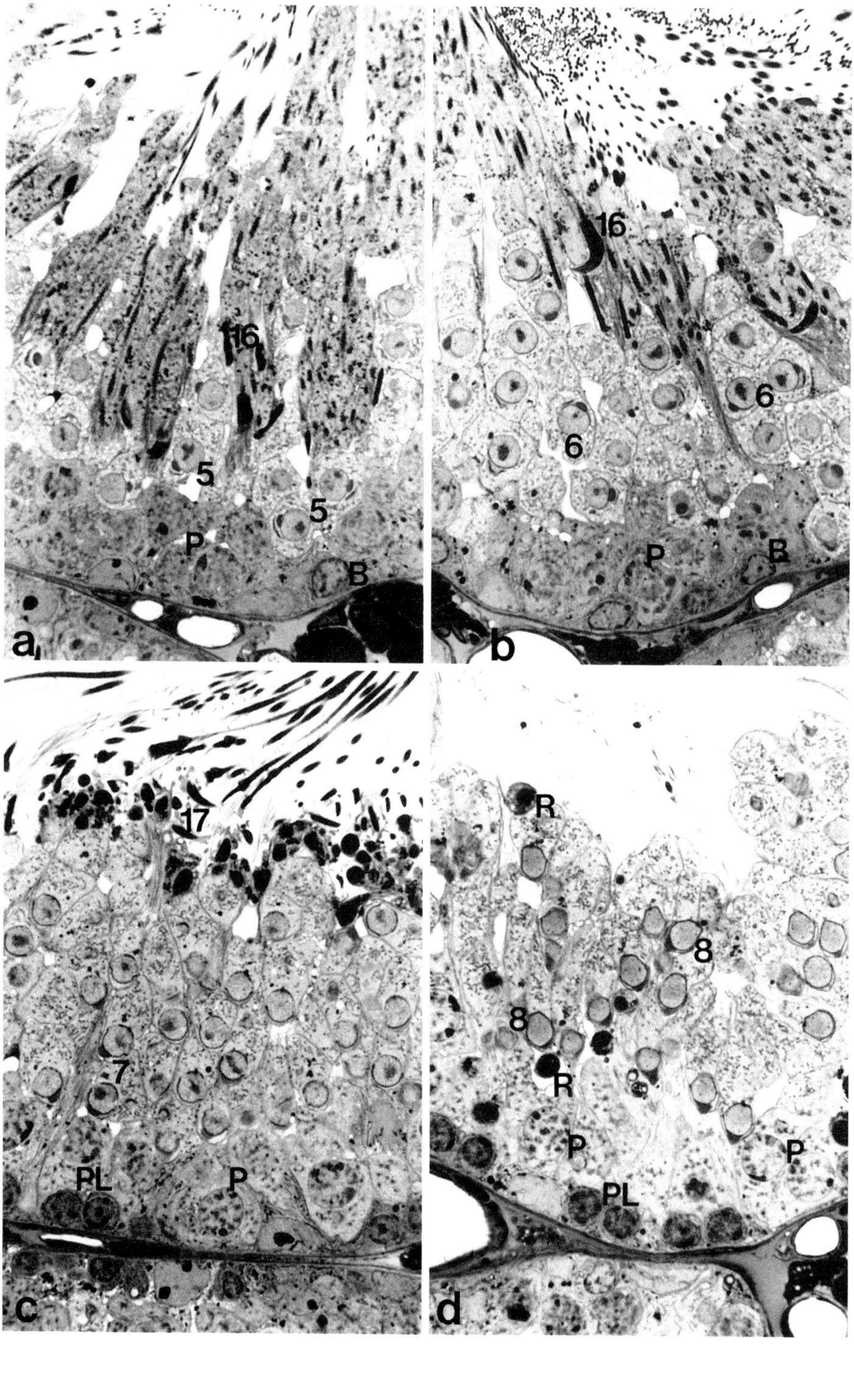

16
5
5
P
B
a
16
6
6
P
B
b
17
7
PL
P
c
R
8
8
R
P
P
PL
d

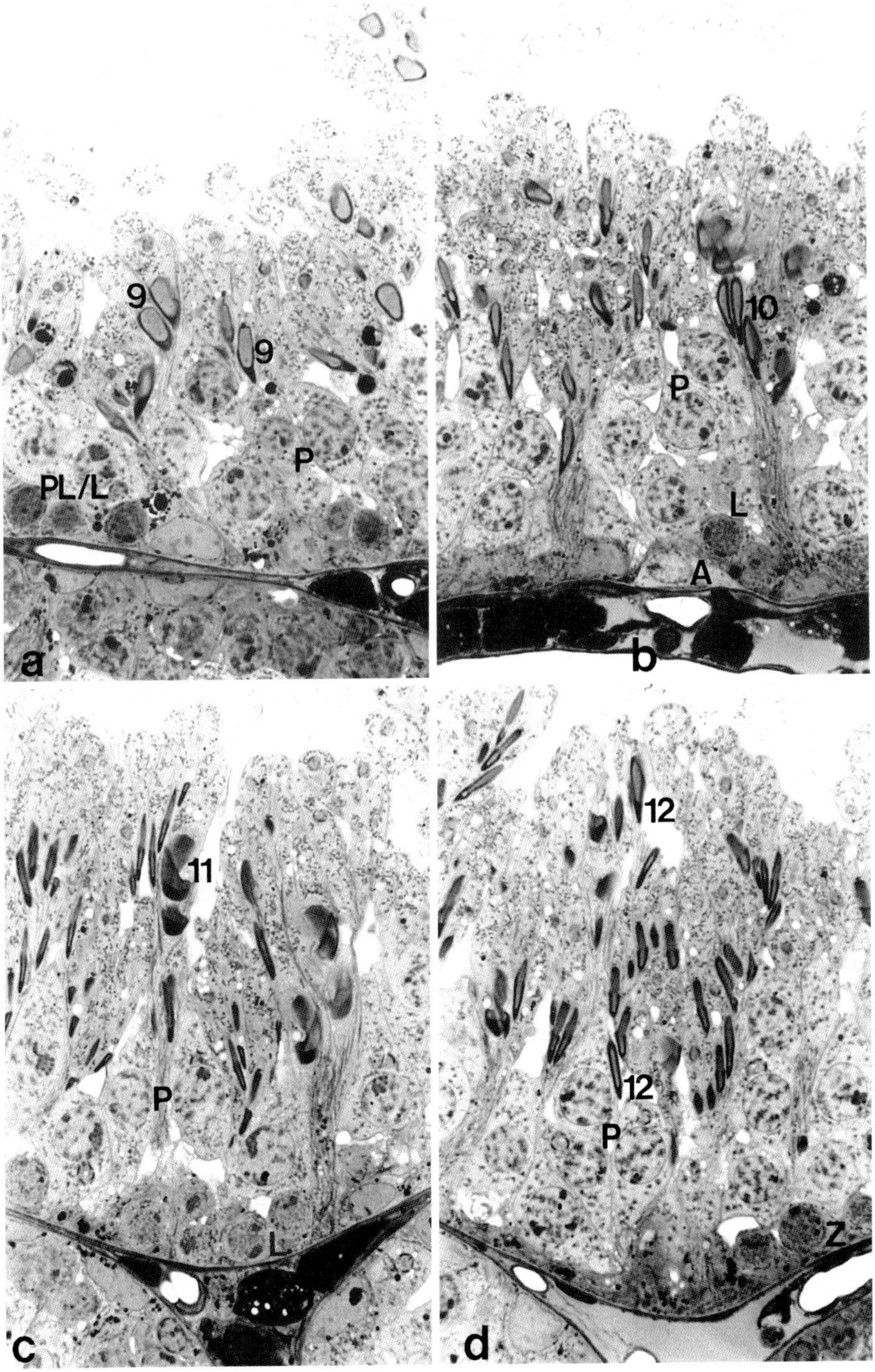
9
9
PL/L
P
a
10
P
L
A
b
11
P
L
c
12
12
P
Z
d

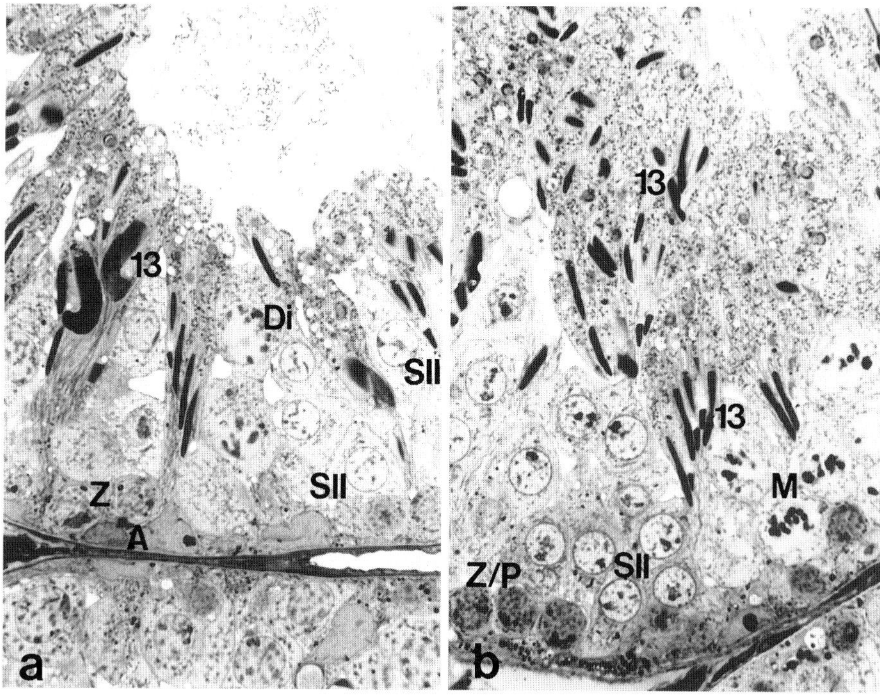

Fig. 6a,b. Light micrographs of stage XIII of adult spermatogenesis. ×700. *A*, A-spermatogonium; *Z*, zygotene spermatocytes; *Z/P*, transition from zygotene to early pachytene spermatocytes; *Di*, diplotene spermatocytes; *SII*, spermatocytes II; *M*, meiotic divisions; *13*, spermatids of step 13

Fig. 5a–d. Light micrographs of **a** stage IX, **b** stage X, **c** stage XI, and **d** stage XII of adult spermatogenesis. ×700. *A*, A-spermatogonium; *PL/L*, transition from preleptotene to leptotene spermatocytes; *L*, leptotene spermatocytes; *Z*, (early) zygotene spermatocytes; *P*, pachytene spermatocytes; *9–12*, spermatids of steps 9–12

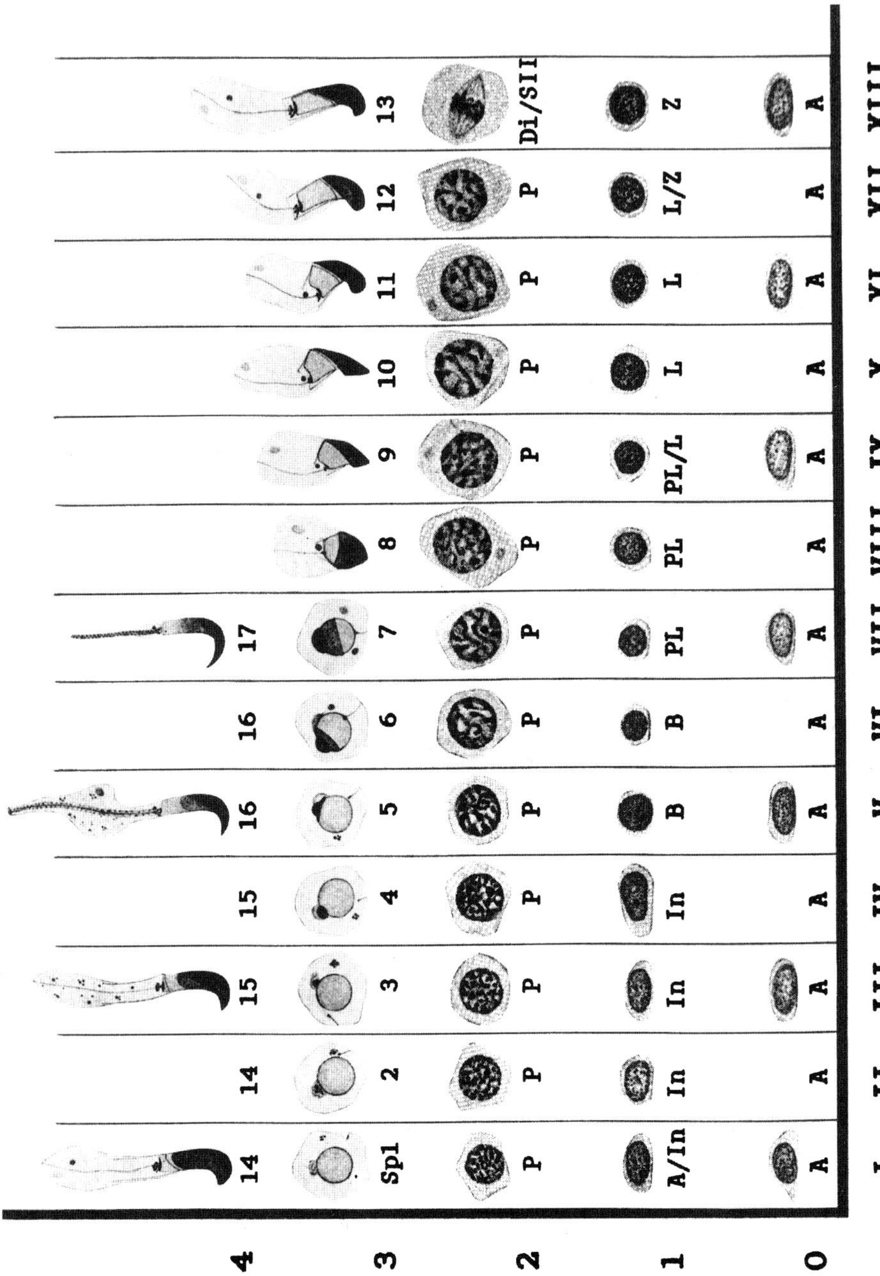

Fig. 7. Cycle map of spermatogenesis shown in Fig. 2 underlaid with a system of coordinates. *I–XIII*, stages I–XIII of spermatogenesis; *A*, A-spermatogonia; *In*, intermediate spermatogonia; *B*, B-spermatogonia; *PL*, postmitotic/preleptotene spermatocytes I; *L*, leptotene spermatocytes I; *Z*, zygotene spermatocytes I; *P*, pachytene spermatocytes I; *Di*, diplotene spermatocytes I; *SII*, spermatocytes II; *Sp1–17*, spermatids of steps 1–17 of spermiogenesis

underlaid with a system of coordinates (Fig. 7), every type of germ cell can be assigned to a pair of coordinate values; the abscissa (Roman numerals I–XIII) indicates the stage or, in the case of cell types persisting for a longer period, a group of stages of the seminiferous epithelium while the ordinate (Arabian numerals 0–4) marks the vertical position within a stage. The pair of coordinates VIII.1, for example, designates preleptotene spermatocytes, while IV.3 designates spermatids at step 4 of spermiogenesis. Based on this nomenclature, any given developmental step of the pubertal seminiferous epithelium is designated in the following by the *pair of coordinate values that applies to the leading (i.e., developmentally most advanced) type of germ cell* present. If the leading type of germ cell is one that is present in more than one stage, coexamination of the associated younger cells allows a more precise determination of the stage (and thus of the abscissa value) applying to such a step of development. It is frequently possible to determine a smaller interval of stages in these cases. For instance, the developmental step of the seminiferous epithelium in which leptotene spermatocytes are the leading germ cells is designated as IX-XII.1, while the step with pachytene spermatocytes leading and associated B-spermatogonia is referred to as V-VI.2. The bottom line of the above system of coordinates (zero) refers to a succession of several generations of A-spermatogonia. Because of their morphological uniformity, these cell types are not considered when determining the developmental step of the seminiferous epithelium; areas showing A-spermatogonia as the only (leading) type of germ cell are designated as step I.1.

3.2.2 Sequence of Appearance

From the appearance of A-spermatogonia in the pubertal seminiferous epithelium (step I.1) onwards up to the complete maturation of the initial generation of spermatids (first spermiation in step VII.4), there is a total of 33 morphologically defined developmental steps of the seminiferous epithelium. The sequence of their appearance follows the same developmental course of germ cells that is also characteristic of adult spermatogenesis. In Table 1, developmental steps and the respective leading germ cells are listed in order of appearance. Figures 8–10 demonstrate exemplarily their appearance in semithin sections, tracing the successive establishment of the seminiferous epithelium up to one of the adult stages IV, VII, and XIII, respectively.

Figure 8 shows steps (I-)IV.1, (I-)IV.2, IV.3, and IV.4. The leading germ cells are In-spermatogonia, (early-)pachytene spermatocytes, spermatids of step 4 and step 15, respectively. In addition to the developmental progress of germ cells, the following alterations are evident: numerically, germ cells come to the fore compared to Sertoli cells; a tubular lumen begins to form between steps IV.1 and IV.2; finally, tubular diameter increases considerably.

Figure 9 shows steps VII(-IX).1, VII(-IX).2, VII.3, and VII.4. The leading types of germ cells are postmitotic (preleptotene) spermatocytes, midpachytene spermatocytes, spermatids of step 7 and step 17, respectively. Progress of germ cells is again accompanied by an increasing numerical ratio of germ cells to Sertoli cells, the formation of a tubular lumen (beginning between steps VII.1 and VII.2), and an increase in tubular diameter. Step VII.4 marks the first occurrence of spermiation and thus represents the end point of the period studied.

Table 1. The 33 developmental steps of the pubertal seminiferous epithelium in the order of their successive appearance, including the leading type of germ cell in each

Developmental step of seminiferous epithelium	Leading type of germ cell (and associated germ cells)
I.1	A-Spermatogonia
I-IV.1	In-Spermatogonia
V-VI.1	B-Spermatogonia
VII-IX.1	Preleptotene spermatocytes
IX-XII.1	Leptotene spermatocytes
XII-XIII.1	Zygotene spermatocytes
I.2	Pachytene spermatocytes (and A-spermatogonia)
I-IV.2	Pachytene spermatocytes (and In-spermatogonia)
V-VI.2	Pachytene spermatocytes (and B-spermatogonia)
VII-IX.2	Pachytene spermatocytes (and preleptotene spermatocytes)
IX-XII.2	Pachytene spermatocytes (and leptotene spermatocytes)
XII-XIII.2	Pachytene spermatocytes (and zygotene spermatocytes)
XIII.2	Diplotene spermatocytes/spermatocytes II/divisions
I.3	Step-1 spermatids
II.3	Step-2 spermatids
III.3	Step-3 spermatids
IV.3	Step-4 spermatids
V.3	Step-5 spermatids
VI.3	Step-6 spermatids
VII.3	Step-7 spermatids
VIII.3	Step-8 spermatids
IX.3	Step-9 spermatids
X.3	Step-10 spermatids
XI.3	Step-11 spermatids
XII.3	Step-12 spermatids
XIII.3	Step-13 spermatids
I.4	Step-14 spermatids (and step-1 spermatids)
II.4	Step-14 spermatids (and step-2 spermatids)
III.4	Step-15 spermatids (and step-3 spermatids)
IV.4	Step-15 spermatids (and step-4 spermatids)
V.4	Step-16 spermatids (and step-5 spermatids)
VI.4	Step-16 spermatids (and step-6 spermatids)
VII.4	Step-17 spermatids/spermiation (and step-7 spermatids)

Fig. 8a–d. Light micrographs of the developmental steps of the seminiferous epithelium representing the successive establishment of adult stage IV. ×700. **a** Step (I-)IV.1, composed of In-spermatogonia (*In*) and Sertoli cells (*S*). **b** Step (I-)IV.2, composed of In-spermatogonia, (early-)pachytene spermatocytes (*P*), and Sertoli cells. **c** Step IV.3, composed of In-spermatogonia, (early-)pachytene spermatocytes, spermatids of step 4 (*4*), and Sertoli cells. **d** Step IV.4, composed of In-spermatogonia, (early-)pachytene spermatocytes, spermatids of steps 4 and 15 (*15*), and Sertoli cells. Note expansion of tubular diameter, lumen formation (beginning between steps (I-)IV.1 and (I-)IV.2), and considerable increase in the numerical germ cell to Sertoli cell ratio

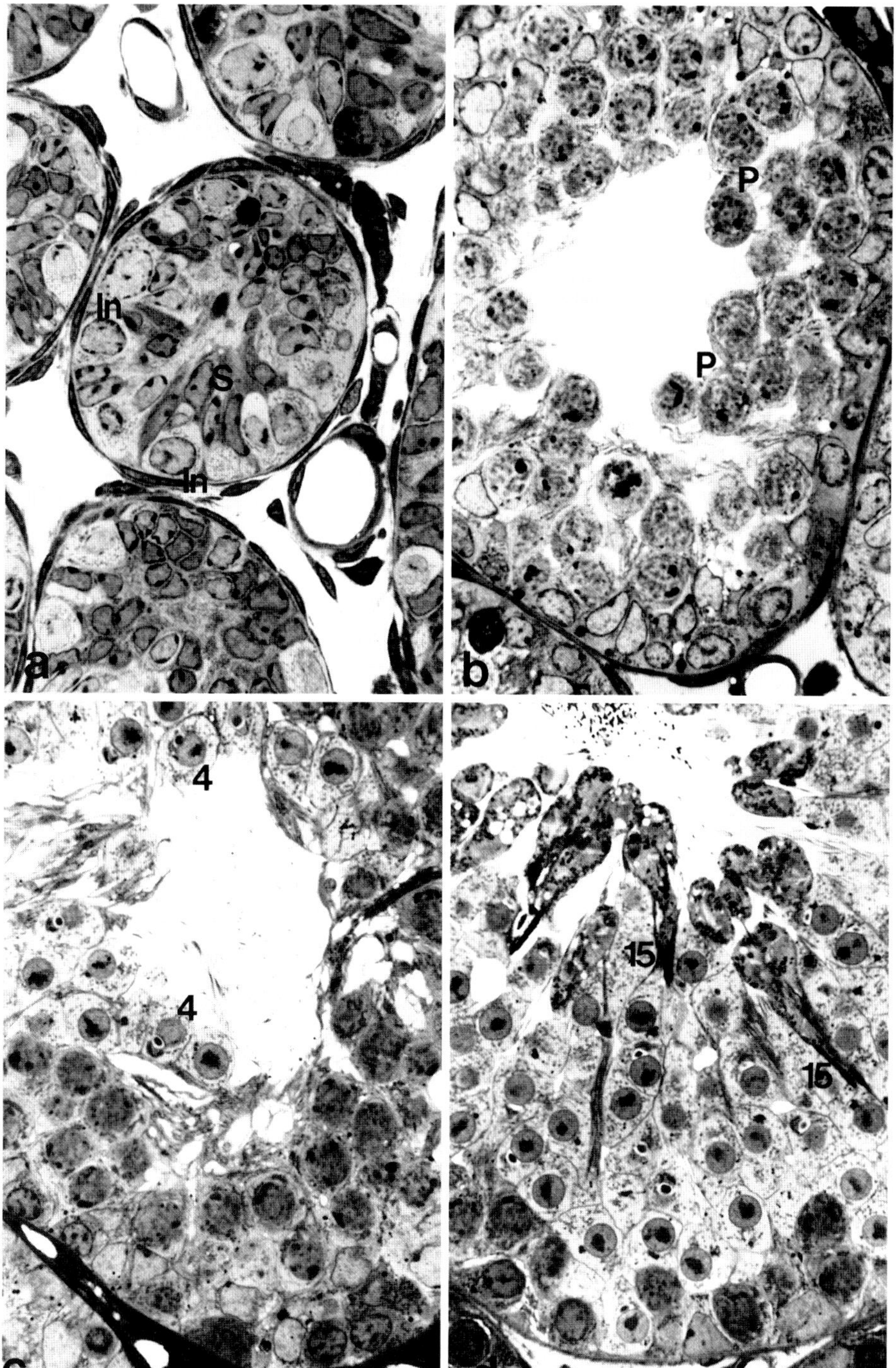

In
S
In
a
P
P
b
4
4
c
15
15
d

Fig. 9a–d. Light micrographs of the developmental steps of the seminiferous epithelium representing the successive establishment of adult stage VII. ×700. **a** Step VII(-IX).1, composed of A-spermatogonia, postmitotic/preleptotene spermatocytes (*PL*), and Sertoli cells (*S*). **b** Step VII(-IX).2, composed of A-spermatogonia, postmitotic/preleptotene and (mid)pachytene (*P*) spermatocytes, and Sertoli cells. **c** Step VII.3, composed of A-spermatogonia, postmitotic/preleptotene and (mid)pachytene spermatocytes, spermatids of step 7 (*7*), and Sertoli cells. **d** Step VII.4, composed of A-spermatogonia, postmitotic/preleptotene and (mid)pachytene spermatocytes, spermatids of steps 7 and 17 (*17*), and Sertoli cells

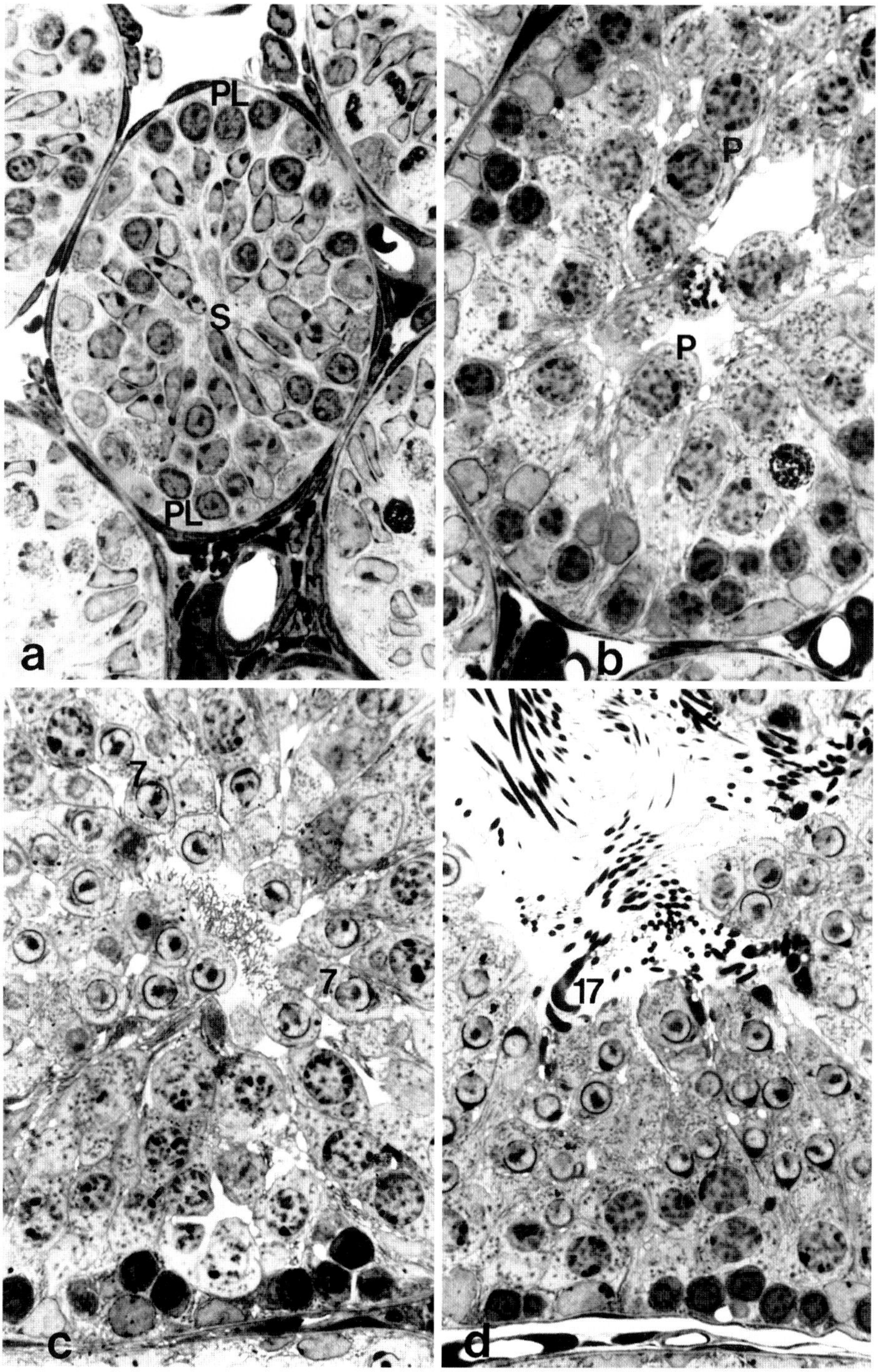
PL
S
PL
a
P
P
b
7
7
c
17
d

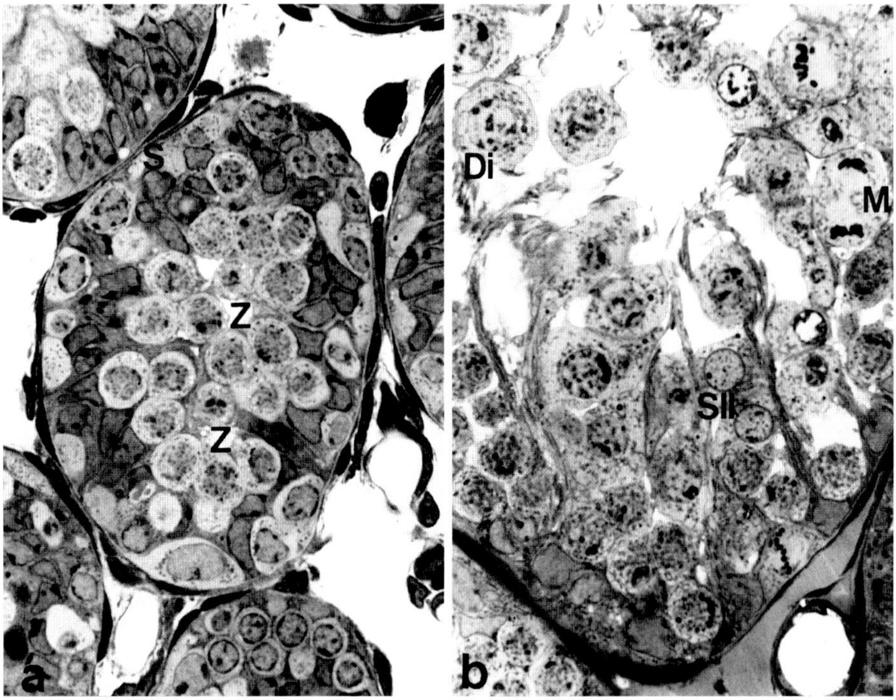

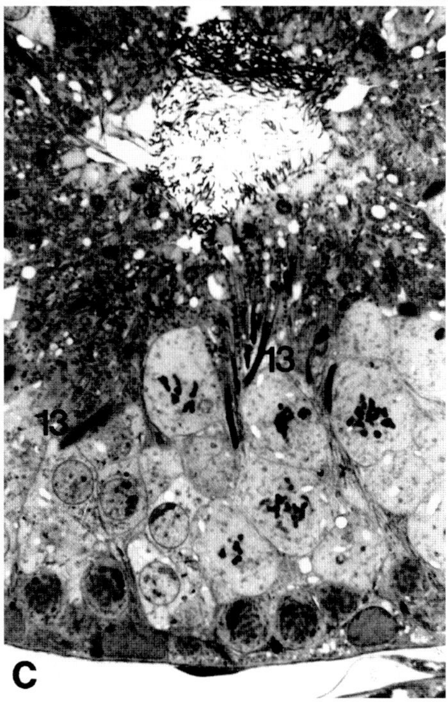

Fig. 10a–c. Light micrographs of the developmental steps of the seminiferous epithelium representing the successive establishment of adult stage XIII. ×700. **a**. Step (XII-)XIII.1, composed of A-spermatogonia, zygotene spermatocytes (*Z*), and Sertoli cells (*S*). **b** Step XIII.2, composed of (dividing) A-spermatogonia, zygotene spermatocytes, diplotene spermatocytes (*Di*), spermatocytes II (*SII*), and meiotic divisions (*M*), and Sertoli cells. **c** Step XIII.3, composed of A-spermatogonia, zygotene spermatocytes, spermatocytes II, meiotic divisions, spermatids of step 13 (*13*), and Sertoli cells

Figure 10 shows developmental steps (XII-)XIII.1, XIII.2, and XIII.3. The leading germ cells are zygotene spermatocytes, diplotene spermatocytes (or interphase secondary spermatocytes or dividing spermatocytes), and spermatids of step 13, respectively. Here, as in the examples above, the number of germ cells increasingly outweighs the number of Sertoli cells. The tubular lumen, which is already discernible at step XIII.1, gradually increases in the subsequent steps, as does the tubular diameter.

3.2.3
Synchronization of Germ Cell Development

In the following, the developmental state of those germ cells that are also present in the respective developmental steps of the seminiferous epithelium in addition to the leading types of germ cells is analyzed. This is first exemplified on the basis of the micrographs referred to in the previous section. Step IV.3 (Fig. 8c) exhibits (early-) pachytene spermatocytes and In-spermatogonia as additional cell types. Step IV.4 (Fig. 8d) shows (early-)pachytene spermatocytes and In-spermatogonia in addition to the leading spermatids of steps 15 and 4. Without exception, these are cells that are also present in adult stage IV of the seminiferous epithelium (compare with Figs. 2, 3d). An analogous situation is found in the other two examples documented above: in both step VII.3 (Fig. 9c) and step VII.4 (Fig. 9d), (mid)pachytene and postmitotic (preleptotene) spermatocytes, which are also found in adult stage VII (see Figs. 2, 4c), are also present. In step XIII.3 (Fig. 10c), diplotene spermatocytes (and/or secondary spermatocytes or spermatocyte divisions) and zygotene spermatocytes, which are also found in adult stage XIII (see Figs. 2, 6), can be seen in addition to the leading spermatids of step 13.

A comprehensive analysis of the germ cell composition of all developmental steps reveals that the demonstrated constancy of germ cell types associated with a given step is a general feature of the pubertal seminiferous epithelium. As in the above examples, the germ cell associations regularly correspond to one of the adult stages of spermatogenesis. This phenomenon shall be referred to as the *stage conformity* of the developmental steps in the following. A single departure from this principle, regarding a certain phase of spermatid maturation, is described in detail in Sect. 3.3.

The following illustration of two steps immediately following one another in the developmental course documents the precisely synchronized progression of associated types of germ cells. Transition from step V-VI.2 (Fig. 11a) to step VII-IX.2 (Fig. 11b) includes development of the leading early-pachytene to midpachytene spermatocytes; concomitantly, the younger germ cell generation changes from B-spermatogonia to postmitotic (preleptotene) primary spermatocytes.

3.2.4
Temporal Appearance

Figure 12 shows the temporal distribution of the above-mentioned developmental steps in the period between days 12 and 38. For each developmental age studied, 50 tubule sections from two animals each were examined and individually assigned to one of the steps of the pubertal seminiferous epithelium listed on the *x*-axis. Assessed

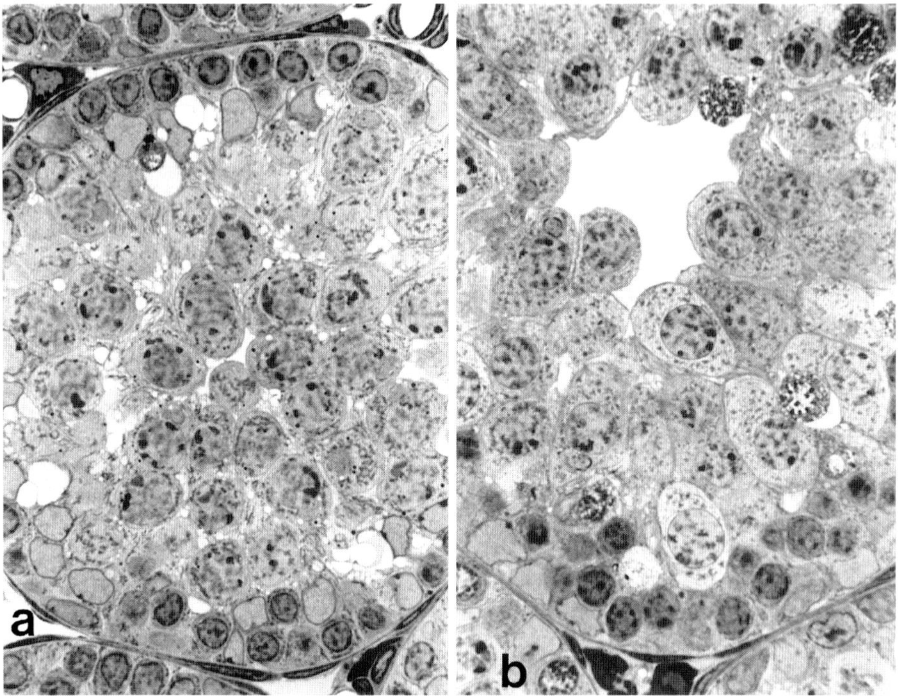

Fig. 11a,b. Progression from **a** step V-VI.2 to **b** step VII-IX.2 shows stage-corresponding development of the germ cell types present: B-spermatogonia advance towards postmitotic/preleptotene spermatocytes, and early-pachytene change to midpachytene spermatocytes. ×700

separately, this results in characteristic sequences of steps present in individual animals and thus represents a measure for the overall developmental state of the respective seminiferous epithelium. For each animal, the sequence of steps is separately plotted as a horizontal bar in Fig. 12. The left end of each bar marks the least advanced and the right end the most advanced developmental step present in a given animal.

3.2.4.1 Duration of Initial Spermatogenesis

The period of time between the initial appearance of A-spermatogonia and the first completion of spermatogenesis (represented by spermiation in step VII.4) lasts from day 12 to day 38. The complete initial course of spermatogenesis thus takes approximately 26 days.

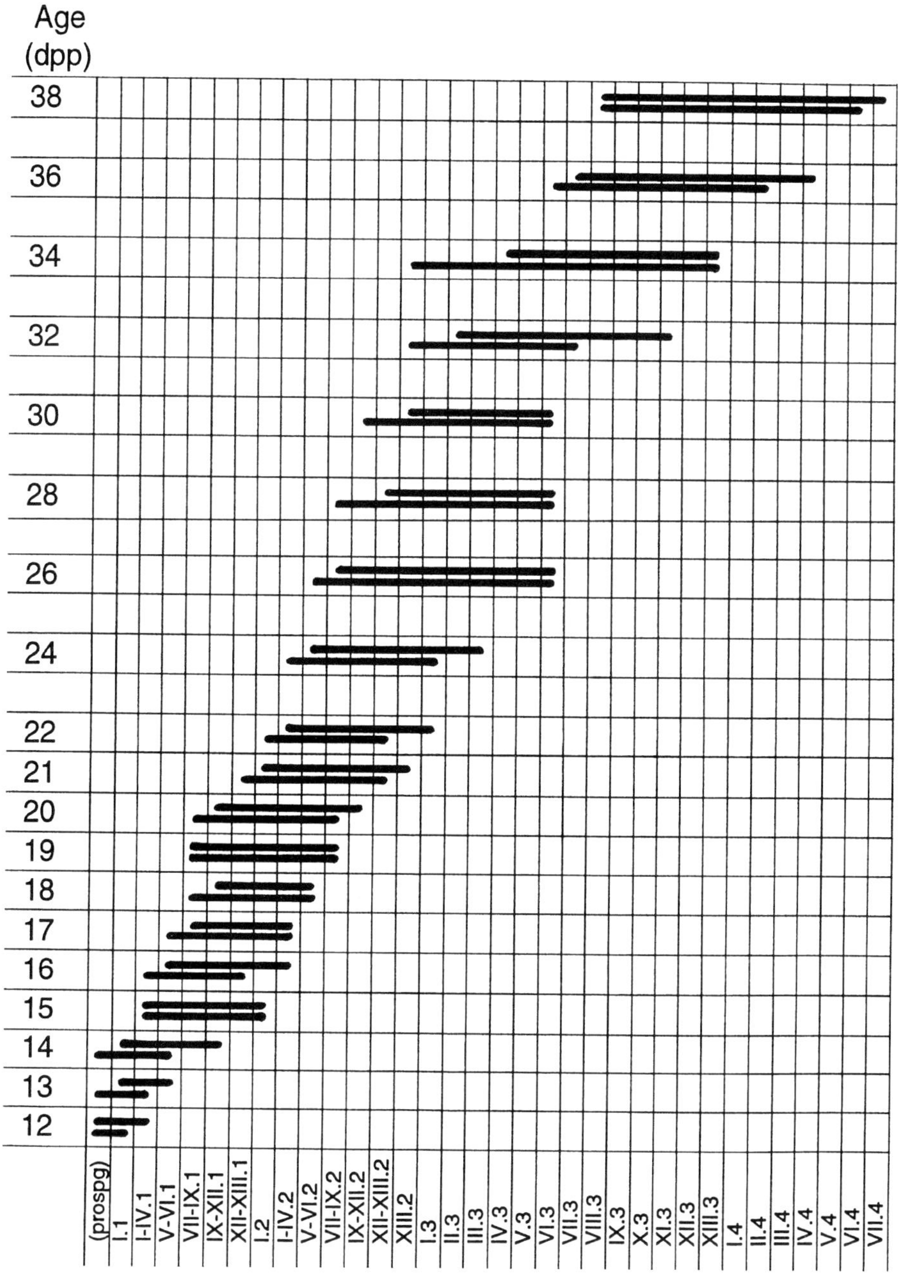

Fig. 12. Survey of the temporal appearance of developmental steps of the seminiferous epithelium in pubertal testes ranging in age from 12 to 38 days. For each of the age levels studied, 50 tubule sections of two animals each were evaluated. *Horizontal bars* represent the developmental interval of steps present in the testis of one individual animal. *dpp*, days post partum; *prospg*, prospermatogonia

3.2.4.2
Intratesticular Variance

The variety of differently developed tubule sections within one and the same testis, already mentioned in Sect. 3.2, can be assessed more precisely on the basis of the data presented in Fig. 12. Each animal studied exhibits a considerable, yet limited variation of the developmental steps present. This phenomenon shall be referred to in the following as the *variance* of the development of the seminiferous epithelium. Its degree is illustrated by the length of the bars in Fig. 12. The absolute bar length, however, is not directly comparable at the different ages, as the subdivision of developmental steps (x-axis) becomes more detailed with increasing age. This results in a gradually increasing length of bars representing developmental intervals of equal length.

By projecting individual intervals of variance onto a schematic map of adult spermatogenesis (Fig. 13), the following relationship becomes evident: throughout all age levels studied, the variance of development determined equals the length of about one line and thus approximately corresponds to the duration of one cycle of the seminiferous epithelium. In Fig. 13, this is illustrated for two ages: on day 16, the intervals of variance in the two animals range from step I-IV.1 up to step XII-XIII.1 and from step V-VI.1 to step I-IV.2, respectively (Fig. 13a); on day 24, these intervals extend from step I-IV.2 to step I.3 and from step V-VI.2 up to step III.3, respectively (Fig. 13b). In all of these cases, the interval covers almost exactly one cycle length of spermatogenesis.

3.2.5
Spatial Arrangement

Any given tubule section shows one and the same developmental step of the seminiferous epithelium throughout its whole circumference. Along with the longitudinal tubular axis, homogeneously developed areas of the seminiferous epithelium are of variable extents and always border on tubule segments belonging to the developmental step immediately preceding or immediately following them. Close examination of individually embedded and longitudinally sectioned parts of tubules reveals the existence of a regular succession of developmental steps along the tubules (Figs. 14, 15). Such continuous sequences of steps exactly corresponds to the developmental interval of one cycle length of spermatogenesis. Within any tubule, the segment showing the least developed and the one showing the most advanced step of the seminiferous epithelium are directly adjacent to one another.

The tubular fragment from a 20-day-old testis demonstrated in Fig. 14 covers the sequence of steps VII-IX.2/VII-IX.1/IX-XII.1/XII-XIII.1/I.2. The gap between the least advanced and the most advanced step of development is situated between steps VII-IX.2 and VII-IX.1. Figure 15 shows the part of a tubule from day 24 containing the succession of steps VII-IX.2/IX-XII.2/XII-XIII.2/XIII.2/I.2/I-IV.2. Here, the developmental gap separates steps XIII.2 and I.2.

Localized reversals of the ascending or descending order of the succession of steps along the tubules is occasionally observed.

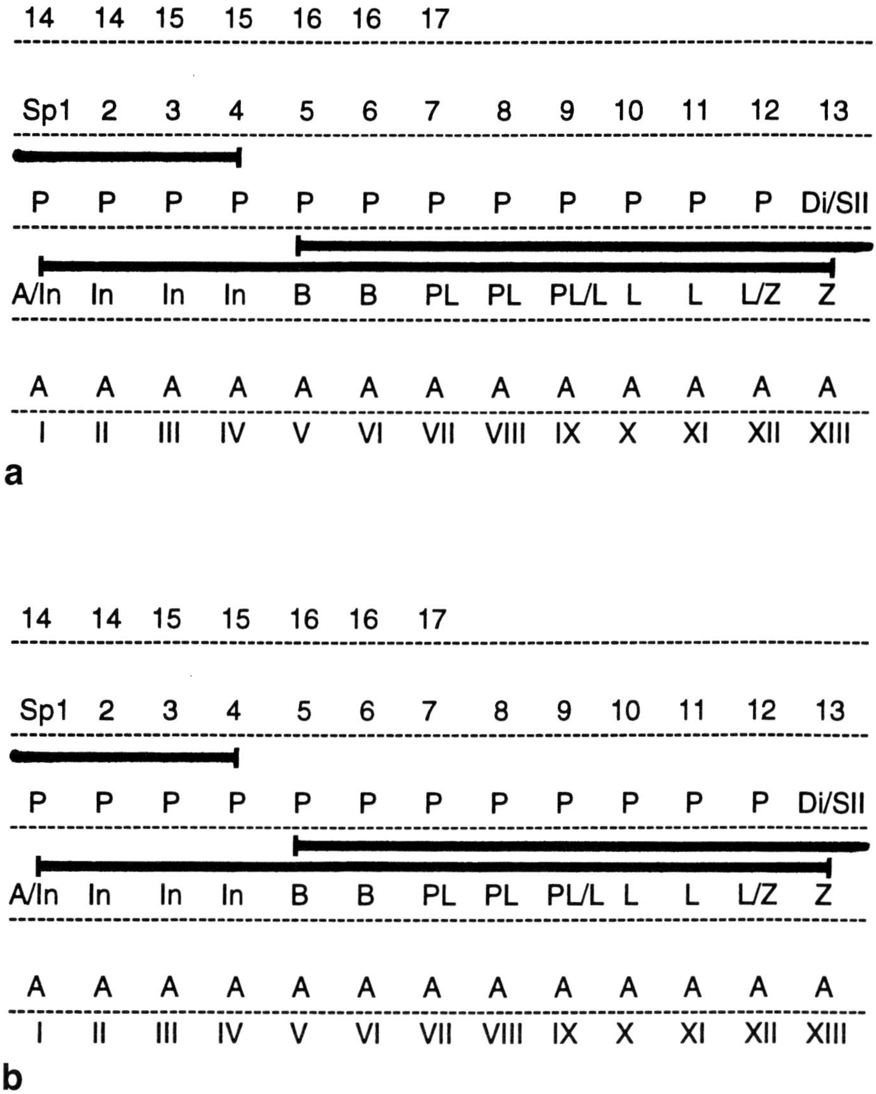

Fig. 13a,b. Exemplary illustration of the intratesticular variance of development of the seminiferous epithelium. **a** On day 16, the animals studied cover a developmental interval extending from step I-IV.1 to step XII-XIII.1 (*lower bar*) and from step V-VI.1 to step I-IV.2 (*upper bar*), respectively. **b** On day 24, animals cover an interval extending from step I-IV.2 to I.3 (*lower bar*) and from step V-VI.2 to III.3 (*upper bar*), respectively. *I–XIII*, stages I–XIII of spermatogenesis; *A*, A-spermatogonia; *In*, intermediate spermatogonia; *B*, B-spermatogonia; *PL*, postmitotic/preleptotene spermatocytes I; *L*, leptotene spermatocytes I; *Z*, zygotene spermatocytes I; *P*, pachytene spermatocytes I; *Di*, diplotene spermatocytes I; *SII*, spermatocytes II; *Sp1–17*, spermatids of steps 1–17 of spermiogenesis

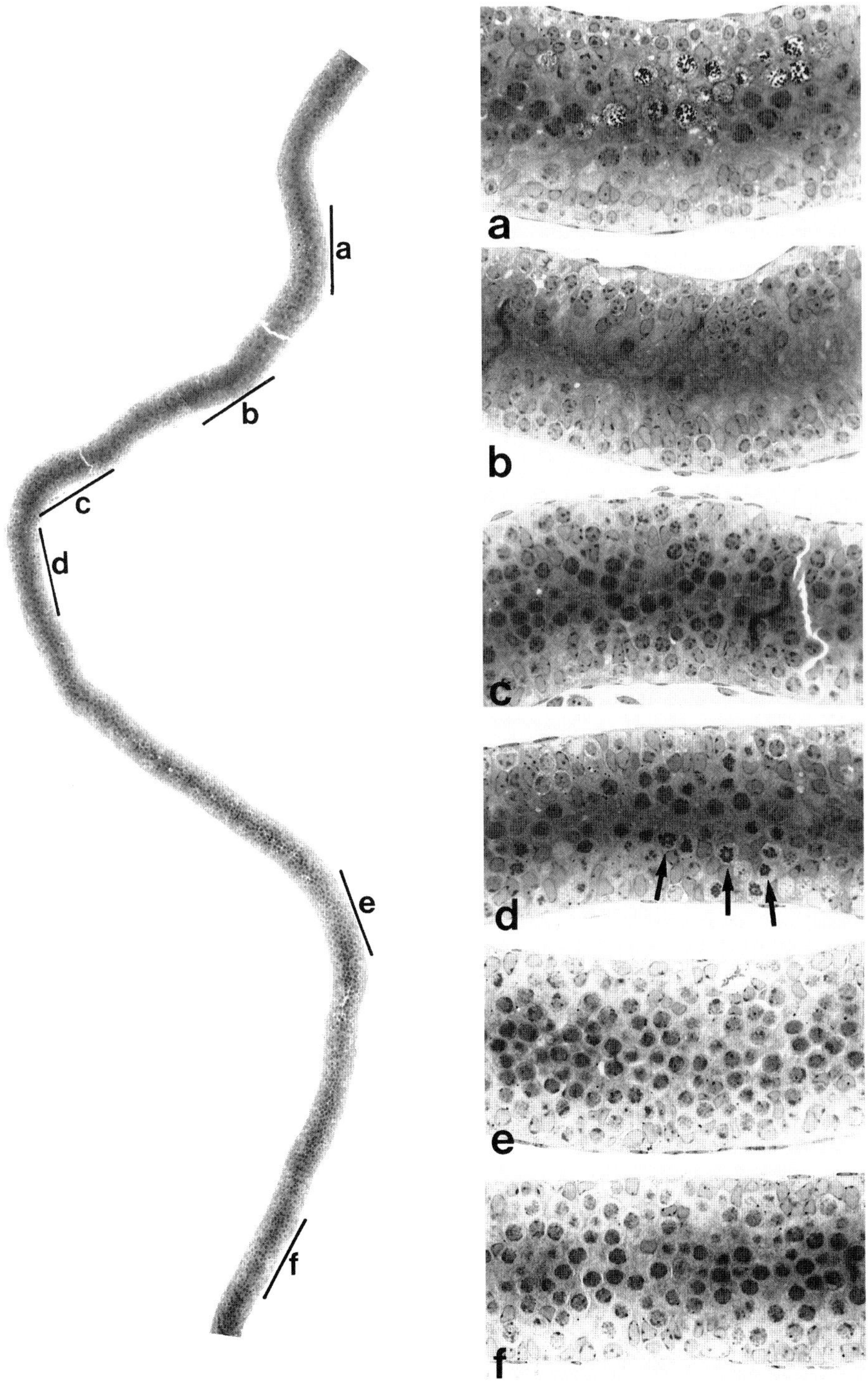
a
b
c
d
e
f

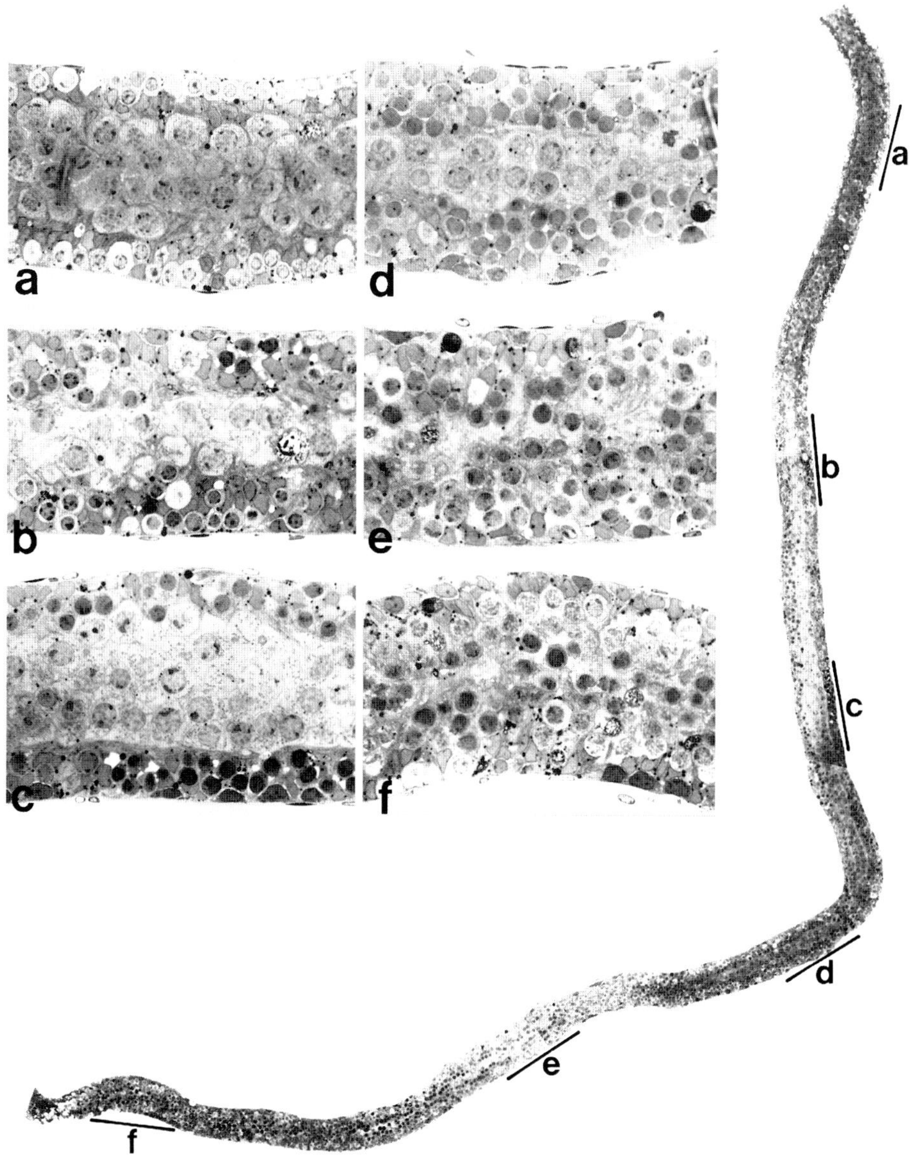

Fig. 15. Longitudinal section of a tubule isolated from a 24-day-old testis. Consecutive developmental steps: **a** Step VII-IX.2. **b** Step IX-XII.2. **c** Step XII-XIII.2. **d** Step XIII.2. **e** Step I.2. **f** Step I-IV.2. *Tubule*, ×60; *details*, ×340

Fig. 14a–f. Longitudinal section of a tubule isolated from a 20-day-old testis. Consecutive developmental steps: **a** Step VII-IX.2, containing degenerating midpachytene spermatocytes (see Sect. 3.6.2). **b** Step VII-IX.1. **c** Step IX-XII.1. **d** Step XII.1, *arrows* indicate specifically arrested germ cell mitoses (see Sect. 3.6.3). **e** Step XII-XIII.1. **f** Step I.2. *Tubule*, ×60; *details*, ×340

3.3 Developmental Delay of Spermiogenesis

3.3.1 Temporal Occurrence

On day 26, the first tubular segments reach step VI.3, with spermatids of step 6 as the leading type of germ cell. However, it is not until after day 30 that segments appear containing spermatids of step 8 and above. In Fig. 12, this phenomenon is reflected in the persistence of the right ends of the bars by the column above step VI.3 between days 26 and 30. At the same time, the matching left ends gradually move to the right, thus indicating an undisturbed progression of germ cells within the developmentally less advanced tubular segments. Beyond day 30, development of the seminiferous epithelium of the leading segments proceeds regularly again.

3.3.2 Morphology

The initial spermatids of step 6 appearing from day 26 onwards are, at first, of regular morphology. The spherical nucleus still occupies a central position within the cell, and the acrosome has spread over almost half of its surface. On days 28 and 30, these spermatids occasionally show a minor developmental progress in that the expanded margin of the acrosome slightly indents the nuclear surface to form a shallow marginal fossa. This, however, does not reach the degree characteristic for spermatids of step 7. In addition, there is an increasing number of degenerative alterations observable in these cells, such as nuclear or acrosomal deformation, bi- or multinucleation, and an occasional premature detachment of the cells into the tubular lumen (see Sect. 3.6.4).

3.3.3 Irregular Germ Cell Associations

On day 28 and, even more frequently, on day 30, tubule sections of step VI.3 appear that do not follow the stage conformity of germ cell associations dealt with in Sect. 3.2.3. Leading germ cells are spermatids of step 6 (or step 6/7; see above). However, the associated younger germ cell generations show a developmental level beyond that characteristic of regular step VI.3. In addition to A-spermatogonia, preleptotene and midpachytene spermatocytes (corresponding to stages VII–IX) or leptotene and late-pachytene spermatocytes (corresponding to stages IX–XII; Fig. 16a) are frequently present. In several instances on day 30, spermatocytes have proceeded up to the zygotene and late-pachytene phase (corresponding to stage XII) or, when present in the zygotene and diplotene phase or as secondary spermatocytes, correspond to stage XIII (Fig. 16b).

Beyond day 30, tubular segments of steps VII.3–XII.3 appear with germ cell associations that, again, do not completely correspond to one of the adult stages of spermatogenesis. The leading spermatids are those of steps 7–12 of spermiogenesis and are associated with younger generations of germ cells which exhibit stage conformity with respect to each other, but which are in each case one or more stages ahead of the leading spermatids. The step X.3 tubule shown in Fig. 16c, for instance, exhibits younger germ cell types (zygotene and late-pachytene spermatocytes) corresponding to stage XII. Within the step XI.3 tubule demonstrated in Fig. 16d, the associated younger cells (A-spermatogonia, early-pachytene spermatocytes, spermatids of step 1) have even reached a level corresponding to stage I.

In addition to the fact that they no longer exhibit stage conformity, such areas of the seminiferous epithelium show other morphological alterations. The members of the leading generation of initially elongating spermatids frequently exhibit a heterogeneous developmental level and thus may not be assigned to one and the same step of spermatid differentiation (Figs. 16c,d). In such cases, the most advanced cells within the leading spermatid generation are considered when determining the developmental step of the seminiferous epithelium. In addition, in many of these tubular segments, the leading spermatids are subject to degenerative alterations of various degrees up to cellular death (see Sect. 3.6.4).

3.4
11β-Hydroxysteroid Dehydrogenase in Pubertal Leydig Cells

3.4.1
Histology of Leydig Cells

Figure 17 illustrates the position and morphology of Leydig cells, which lie singly or in clusters within the loose connective tissue filling the angular interstitial spaces between seminiferous tubules. They are usually associated with capillaries (Fig. 17a) and are often located close to the lamina propria of tubules (Fig. 17a,b).

3.4.2 Demonstration of 11β-Hydroxysteroid Dehydrogenase

The activity of Leydig cell 11β-HSD was semiquantitatively recorded throughout the second half of the developmental period studied, i.e., from day 24 onwards. Tentative reactions in sections of younger testes turned out negative. On day 24, the vast majority of Leydig cells were 11β-HSD negative, and only very sporadically did cells show a weak trace of diffusely distributed staining granules within the cytoplasm (Fig. 18a). On days 26 and 28, several Leydig cells also remained negative; however, an increasing number of cells exhibited a stronger 11β-HSD reaction (Fig. 18b). On day 30, all cells were 11β-HSD positive, but differences were observed regarding staining intensity (Fig. 18c). Nearly all of the Leydig cells had a strong reaction on day 32; on the following days 34–38, all cells showed a generally strong, adult-like reaction within the cytoplasm (Fig. 18d). By comparing the micrographs shown in Fig. 19a,b, the steep increase in this Leydig cell enzyme during the period from day 26 (Fig. 19a) up to day 34 (Fig. 19b) can be seen.

Intercellular variations of staining intensity are observed only during the phase of steeply rising values (days 26–32), but no longer appear from day 34 onwards. While

Fig. 16a–d. Light micrographs of non-stage corresponding developmental steps. ×700. **a** Step VI.3, showing developmentally arrested spermatids of step 6 in the apical part of the epithelium. Associated younger germ cell types (A-spermatogonia, leptotene and pachytene spermatocytes) conform to adult stages IX–XII. Day 28. **b** Step VI.3, showing spermatids of step 6, while associated germ cells (A-spermatogonia, zygotene spermatocytes, spermatocytes II, and meiotic divisions) conform to stage XIII. Day 30. **c** Step X.3, exhibiting heterogeneously differentiated spermatids of steps 8–10 in the apical half of the epithelium. Associated germ cells (A-spermatogonia, zygotene and late-pachytene spermatocytes) conform to stage XII. Day 32. **d** Step XI.3, exhibiting spermatids of steps 9–11. The other germ cell types present (A-spermatogonia, early-pachytene spermatocytes, and spermatids of step 1) conform to stage I. Day 32

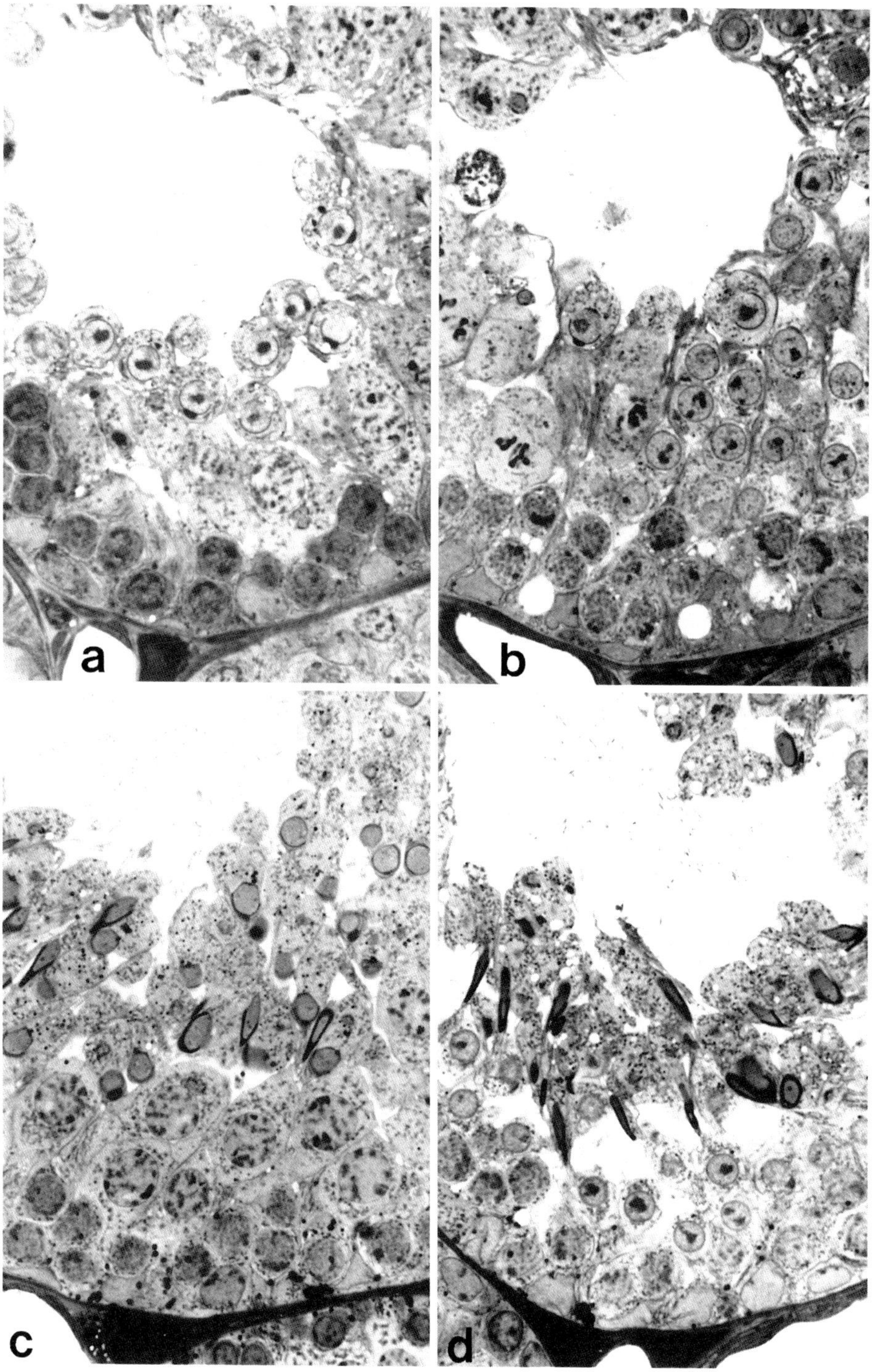

a
b
c
d

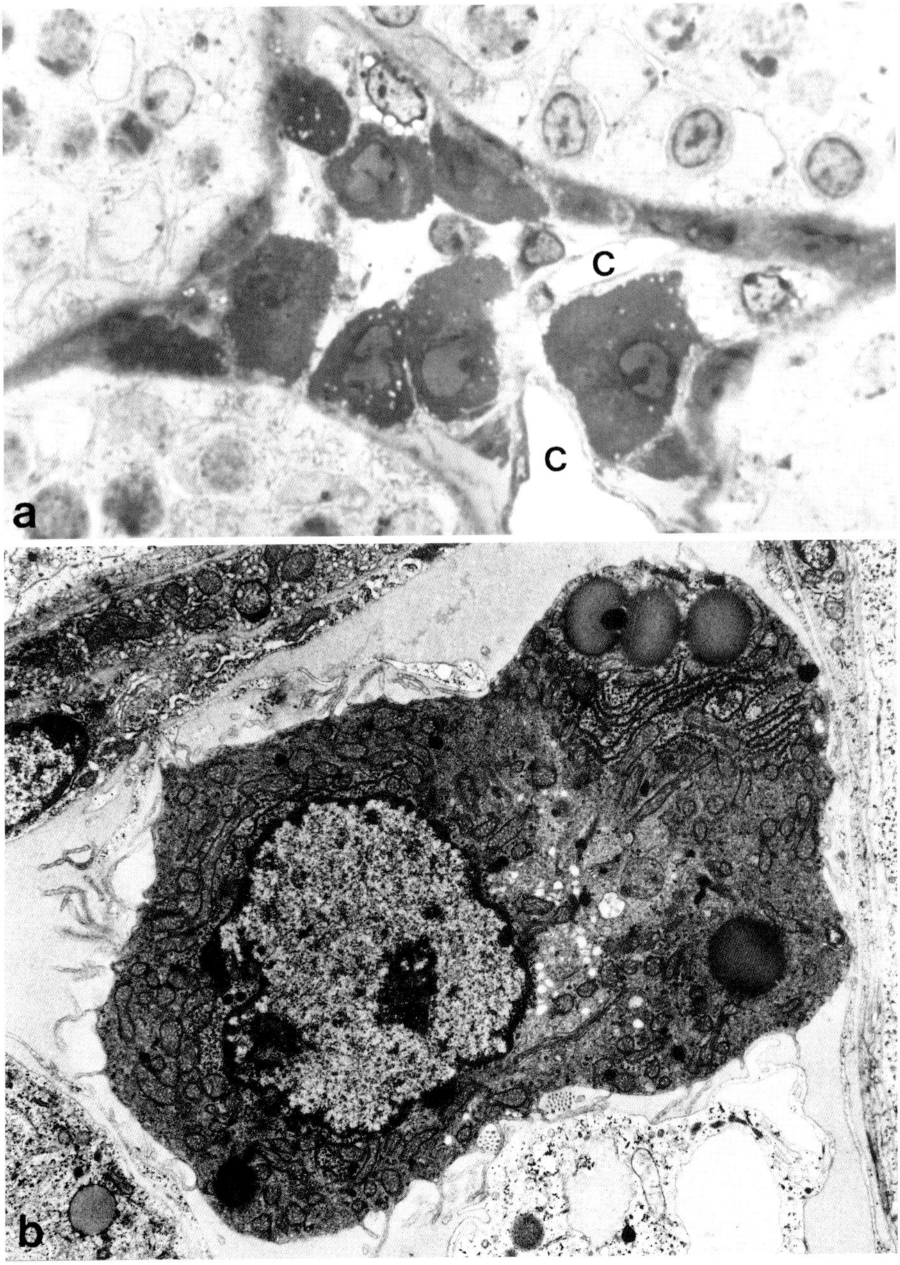

Fig. 17a,b. Position and morphology of Leydig cells, day 28. **a** Light micrograph of a group of Leydig cells lying in close proximity to capillaries (*C*) and to the lamina propria of seminiferous tubules. ×1150. **b** Ultrastructural aspect of a Leydig cell showing mitochondria, cisternae of the endoplasmic reticulum, Golgi cisternae, free ribosomes, and several lipid droplets within the overall darkly contrasted cytoplasm. ×7250

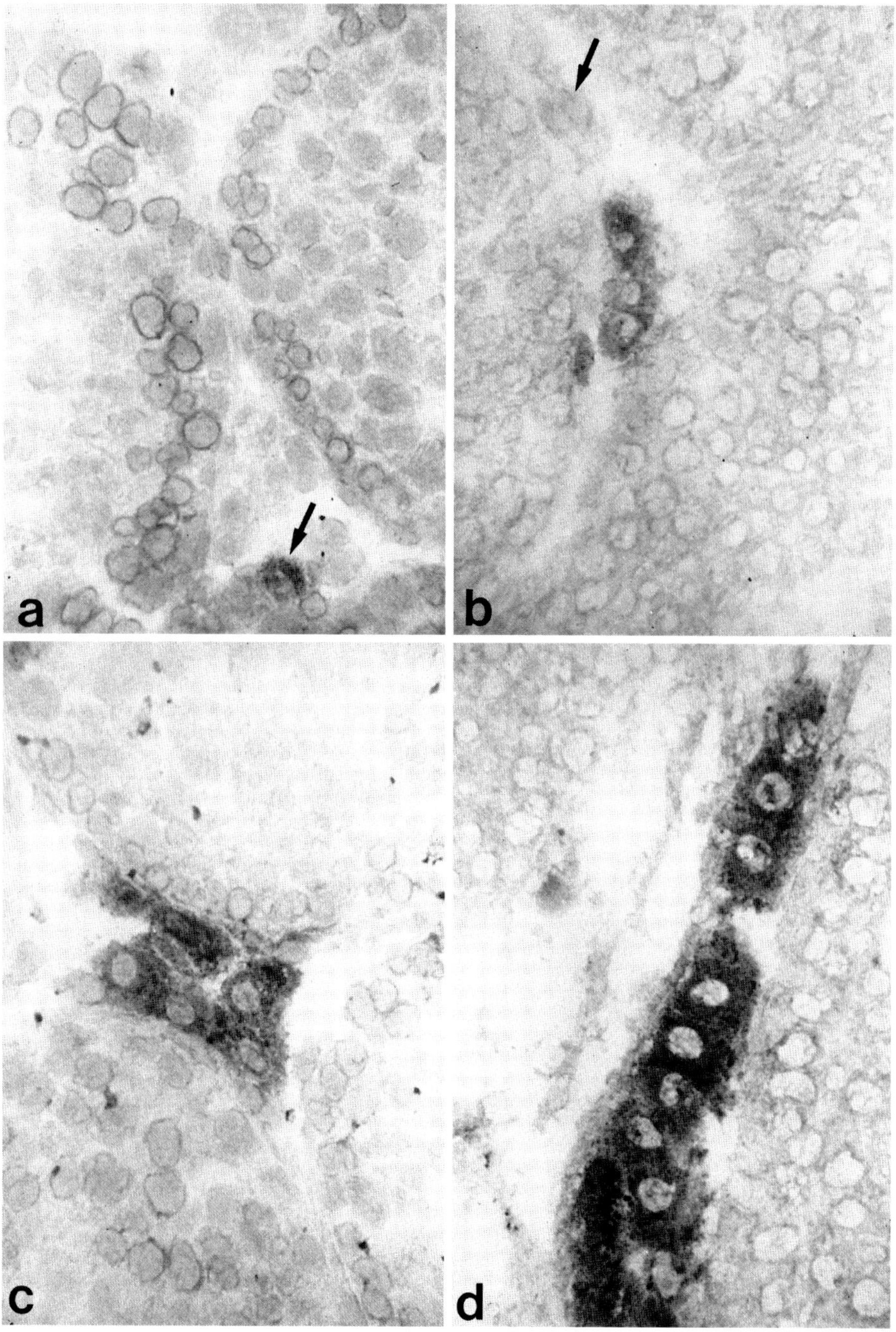

Fig. 18a–d. Enzyme histochemical demonstration of 11β-hydroxysteroid dehydrogenase (11β-HSD) in the Leydig cell cytoplasm at different ages. Slight counterstaining allows recognition of 11β-HSD-negative cells and of adjacent tubules. ×560. **a** Day 24; *arrow* points to one of the weakly stained cells. **b** Day 28; some Leydig cells are still 11β-HSD negative (*arrow*). **c** Day 30. **d** Day 36

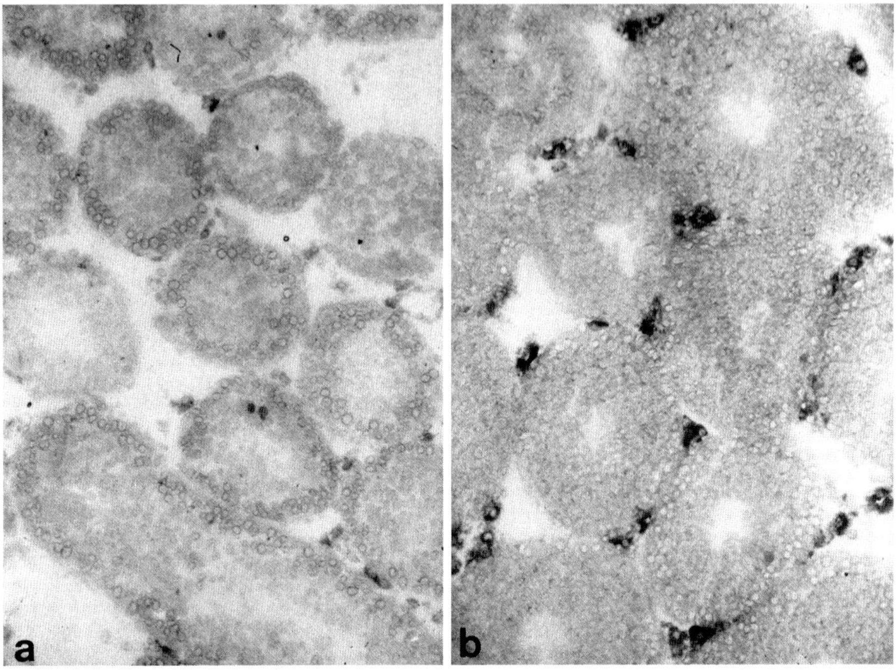

Fig. 19a,b. Survey of 11β-hydroxysteroid dehydrogenase (11β-HSD) establishment in the Leydig cells. Comparative demonstration of 11β-HSD reaction in testicular tissue **a** on day 26 (in parts weakly positive) and **b** on day 34 (strongly positive in all Leydig cells). ×130

staining granules are evenly distributed within the cytoplasm up to day 30, coarse or nearly ring-shaped, irregularly distributed granules are occasionally found on the days that follow (Fig. 18d). Table 2 presents a tabular synopsis of the histochemical results for this enzyme from day 24 to day 38.

3.5 Formation of the Sertoli Cell Barrier

3.5.1 Morphology

From the beginning of the period studied, Sertoli cells are linked by focal membrane contacts both to one another and to adjacent germ cells. The existence of two neighboring Sertoli cell contact zones facing different germ cells (Fig. 20a,b) may lead to an indirect connection of germ cells via a slender lamella of Sertoli cell cytoplasm.

Establishment of the specific contact zones of the Sertoli cell barrier starts off along the basolateral cell membranes. Vesicles and cisternae of the endoplasmic reticulum

Table 2. Semiquantitative and qualitative evaluation of 11β-hydroxysteroid dehydrogenase (11β-HSD) reactions between days 24 and 38

Age (dpp)	Staining intensity[a]	Intercellular differences of staining intensity	Coarsely staining granules
24	0–(+)	0	0
26	0–+	(+)	0
28	0–++	+	0
30	++–++++	+	0
32	+++–++++	+	+
34	++++	0	+
36	++++	0	+
38	++++	0	+

For each of the age levels studied, four cryostat sections from two animals each were included.
dpp, days post partum.
[a]Gradation of staining intensity ranging from 0 (negative) to ++++ (strongly positive, adult-like).

are arranged submembraneously in adjacent cells (Fig. 20c), and numerous tight junctions form in these areas (Fig. 20d). Subsequently, amorphous material and bundles of actin filaments assemble within the cytoplasmic zone separating the subsurface cisternae and the apposed cell membranes (Figs. 20e,f). When tangentially sectioned, actin bundles show a parallel alignment. The total of the basolateral contact zones, including the associated filaments and subsurface cisternae, represents the morphological correlate of the Sertoli cell barrier.

3.5.2 Permeability to Lanthanum

The electron-dense tracer lanthanum, applied together with the fixative, may spread throughout the intercellular spaces of the epithelium as long as no functioning Sertoli cell barrier is present. The epithelial areas of a step XII-XIII.1 tubule (Fig. 21a,b), however, display the tracer exclusively in the basal part of the seminiferous epithelium. At several points, the darkly contrasted lanthanum line abruptly terminates and thus indicates the permeability barrier located above (Fig. 21a). The zygotene spermatocytes are not reached by the tracer. In Fig. 21b, a prominent Sertoli cell contact zone with parallely arranged subsurface bundles of actin filaments is situated immediately above where the lanthanum precipitate ends.

Fig. 20a–f. Morphology of Sertoli cell membrane contacts, day 18. **a** Adjacent areas of membrane contacts (*arrow*) appear to connect preleptotene spermatocytes via a slender Sertoli cell lamella. *Arrowhead*, intercellular bridge; *asterisks*, lamina propria of the tubule. ×4150. **b** Higher magnification of the contact zone illustrates its desmosome-like morphology. ×27,100. **c–f** Successive formation of the specific membrane contacts of the Sertoli cell barrier: **c** Submembraneous arrangement of long cisternae of the endoplasmic reticulum in the cytoplasm of adjoining Sertoli cells. ×38,550. **d** Formation of focal tight junctions (*arrowheads*). ×27,000. **e** Transversely (*arrows*) and obliquely sectioned bundles of actin filaments beneath Sertoli cell membranes. ×54,600. **f** Longitudinally sectioned actin filament bundles (*arrows*) between cell membranes and subsurface cisternae. ×54,600

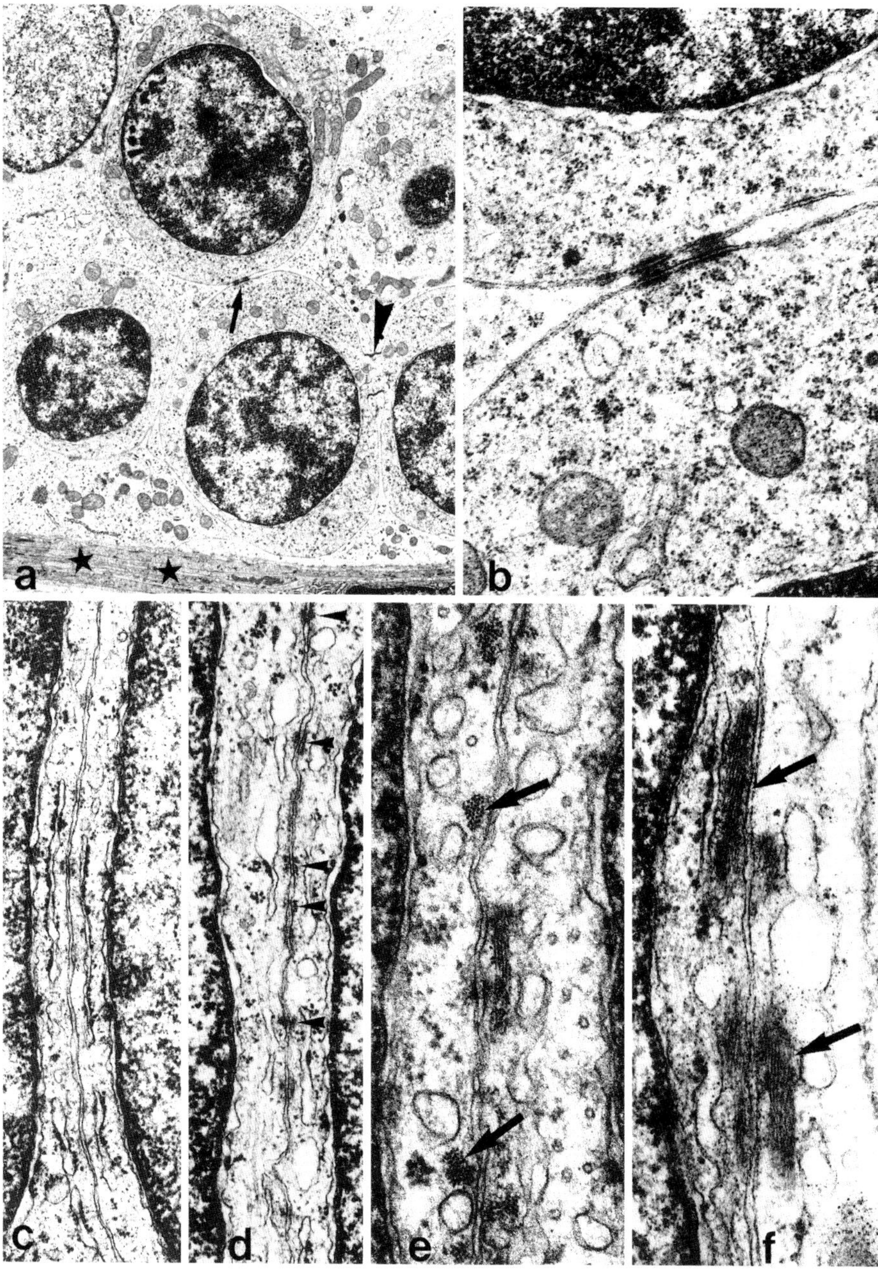

Fig. 21a,b. Demonstration of Sertoli cell barrier function by means of lanthanum. Tubule of developmental step XII-XIII.1, day 19. **a** Lanthanum lines filling the intercellular spaces within the basal area of the epithelium abruptly terminate at several points (*arrows*), indicating the Sertoli cell barrier. Intercellular space surrounding the zygotene spermatocyte (*Z*) is free of the tracer. ×5200. **b** Enlarged detail from the same tubule. Lanthanum stops immediately underneath a prominent contact zone of the Sertoli cell barrier. *Arrowheads* indicate several contact-associated bundles of subsurface actin filaments. ×16,800

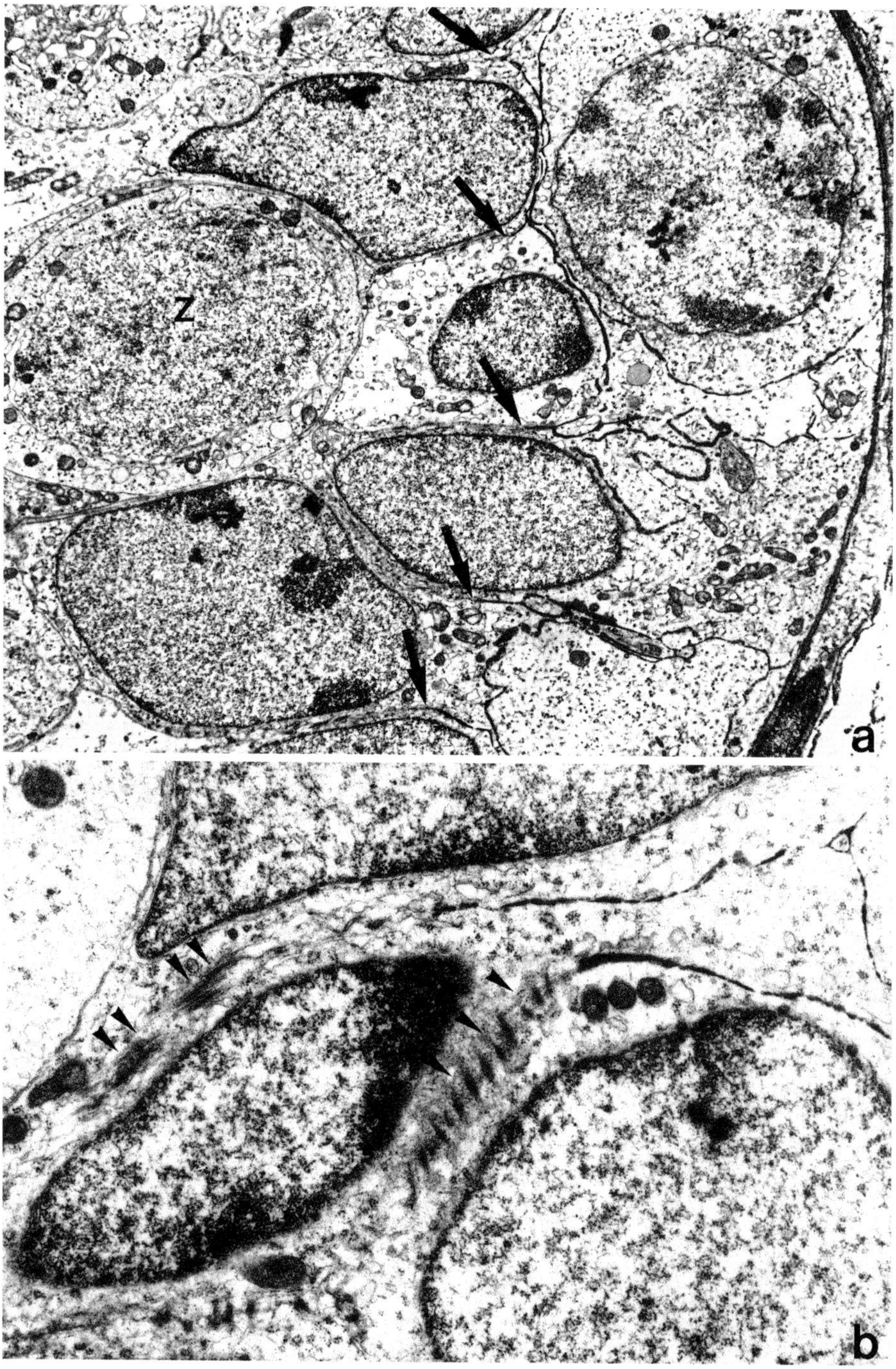
Z
a
b

3.5.3 Synchronization with Germ Cell Development

Morphological establishment and functional maturation of the Sertoli cell contacts is coupled to the developmental progress of the neighboring germ cells. No barrier is present up to and including step VII-IX.1 of the seminiferous epithelium. The barrier forms during steps IX-XII.1 and XII-XIII.1 subjacent to the leading leptotene and zygotene spermatocytes and has been shown to be continuous from step I.2 onwards. Thus spermatocytes are not yet separated by the barrier during the preleptotene phase and are partly separated in the leptotene and zygotene phase. When the leading spermatocytes initially enter the pachytene phase, the Sertoli cell barrier is completed, and from then on compartmentalization of the seminiferous epithelium is maintained continuously.

3.6 Degeneration of Germ Cells

Degenerative germ cells observed during pubertal development of the seminiferous epithelium display a heterogeneous morphological appearance. Various generations of germ cells may be affected, and both the structural course of degeneration and its site and time of occurrence may differ. The main degenerative patterns of germ cells encountered in the present material are described below.

3.6.1 Apoptosis

Numerous degenerating germ cells show key morphological features of apoptosis, such as condensation of both chromatin aggregates along the nuclear membrane and of the cytoplasm, crowding of cell organelles, and lobulation of the nucleus (Fig. 22a). However, both nuclear fragmentation and surface blebbing or splitting of the cell into smaller apoptotic bodies is only occasionally observed. Apoptotic germ cells are regularly phagocytized by neighboring Sertoli cells and are subsequently subjected to a rapidly progressing structural degradation (Fig. 22b).

The apoptotic mode of cell death affects germ cells during various developmental steps of the seminiferous epithelium and does not show a significant temporal accumulation. Cells or groups of cells involved are mainly situated within the basal compartment of the epithelium close to spermatogonia.

3.6.2 Spermatocytes

Spermatocytes frequently degenerate in the course of the meiotic divisions. Morphological signs of degeneration are cytoplasmic vacuolization and deposition of granular material within the cytoplasm, along membranes, and, when present, along the meiotic spindle fibers (Fig. 23). Cell organelles and chromatin structures progressively con-

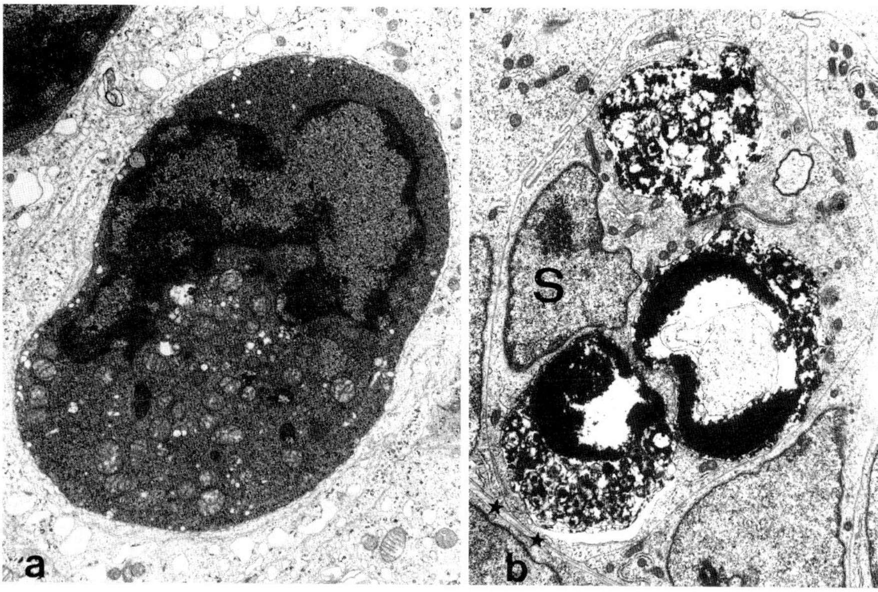

Fig. 22a,b. Germ cell apoptoses. **a** Overall darkly staining apoptotic germ cell showing chromatin aggregation along nuclear membrane, lobulation of nucleus, and crowding of cell organelles opposite the nucleus. Day 16. ×6700. **b** Sertoli cell (its nucleus is labeled with *S*) containing phagozytized remnants of three apoptotic germ cells. *Asterisks* indicate basement membrane of seminiferous epithelium. Day 12. ×5500

dense and the whole cell finally disintegrates, leaving large vacuoles within the seminiferous epithelium. Remnants of degenerating spermatocytes are shed into the tubular lumen rather than being phagocytized by Sertoli cells.

All subphases of meiosis may be affected, but those most often concerned are midpachytene cells (Figs. 23a,b) and metaphases (Figs. 23c,d). Degenerating midpachytene spermatocytes are mainly found in steps VII-IX.2 (Fig. 23a; see also Fig. 14a), VII.3-IX.3, and VII.4. They do not appear prior to day 19. Degenerating metaphases (of meiosis I or II) appear in developmental steps XIII.2 (Fig. 23c) and XIII.3 and are occasionally also encountered in steps I.3 and I.4.

3.6.3 Arrested Mitoses

3.6.3.1 Morphology

Many of the germ cells affected by mitotic arrest at first sight resemble regular metaphase figures. However, the chromosomes form maximally condensed aggregates which are frequently arranged in a rosette-like formation within the cytoplasm (Fig. 24a; see also Fig. 14d). All centrioles assemble in a central position inside the circle

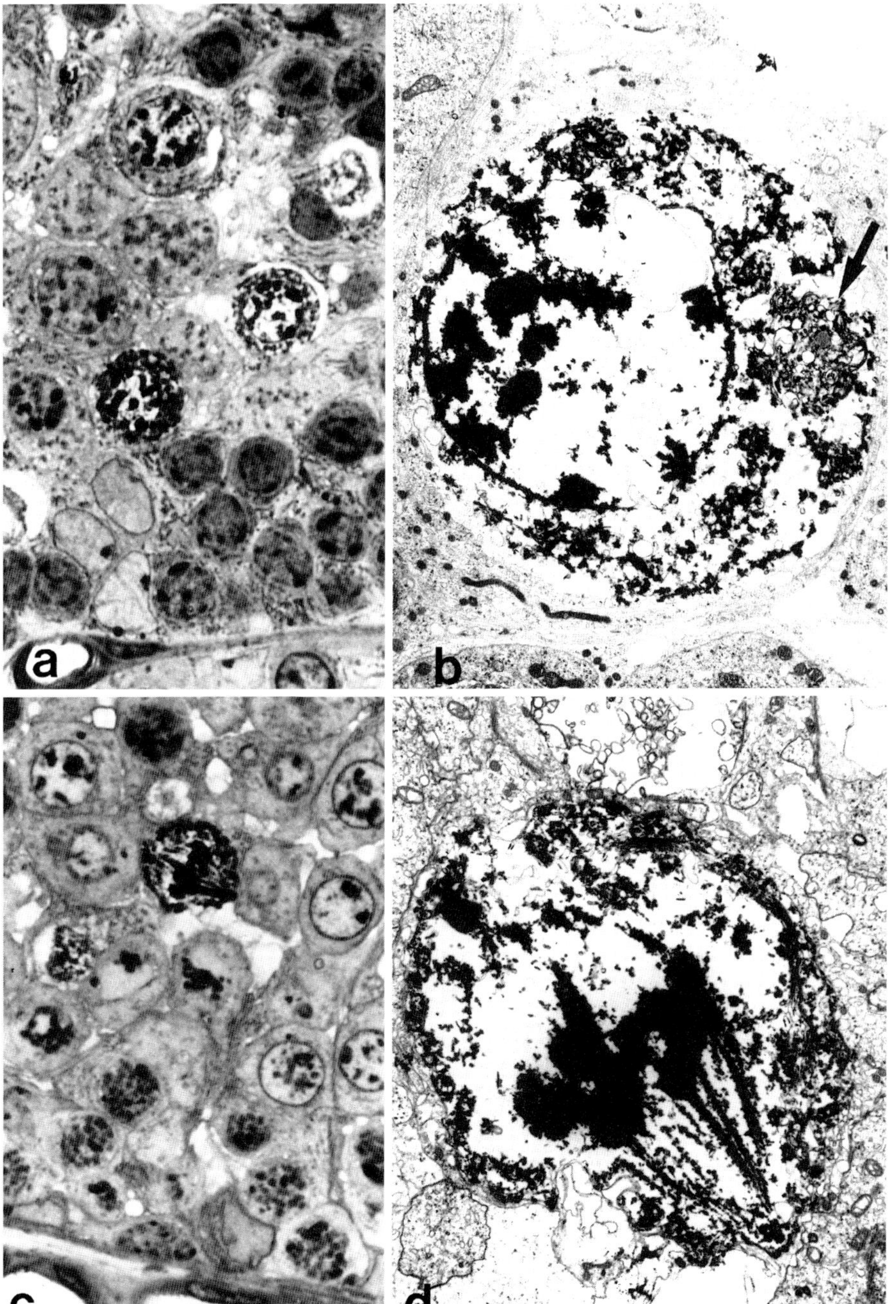
a
b
c
d

of chromosomes (Fig. 25a). Starting from the pericentriolar area, irregular microtubules may proceed in a radial direction towards the chromosomes or between them. The cells involved lie at an intermediate level of the epithelium among spermatocytes or are located more basally close to (but without immediate contact with) the basement membrane of the tubule. They frequently appear as groups of cells probably representing one or several germ cell clones. Neighboring cells are joined by open intercellular bridges (Fig. 25b), the lumen of which may be occupied by various components of the meiotic apparatus, such as spindle fibers and centrioles, or parts of chromosomes (Fig. 25b).

3.6.3.2 Formation and Degenerative Fate

Development towards the above-mentioned morphological appearance begins with a gradually increasing condensation of chromosomes, a process which may be in different stages within the nucleus of one cell (Fig. 24b, 26a). Maximal chromosomal condensation is not achieved until the nuclear membrane has disappeared. The intermediate degenerative phases depicted in the light micrograph shown in Fig. 24b are in the process of disassemblance of the nuclear envelope. Ultrastructural analysis of one of these cells reveals that several centrioles have already reached the characteristic central position among the chromosomes, while parts of the perinuclear cistern are still visible (Fig. 26b).

After persisting in the above state for a certain time, shrinkage of the whole cell and increasing compression of chromosomes and cytoplasmic structures indicate the final destruction of the cells (Fig. 24c).

3.6.3.3 Spatial and Temporal Appearance

Arrested germ cell mitoses appear exclusively in developmental steps VII-IX.1 to XII-XIII.1. They are only found in those tubule segments in which the neighboring spermatocytes have reached the preleptotene, leptotene, or zygotene phase of meiosis I for the first time (Figs. 24a–c, 14d). Accordingly, occurrence of this mode of germ cell degeneration is confined to the period between days 14 and 20. Table 3 summarizes the appearance of these cells both in relation to the developmental state of the local seminiferous epithelium and in relation to the ages involved. The center of distribution is located in those developmental steps in which spermatocytes are in transition from

◄

Fig. 23a–d. Degenerating spermatocytes. **a** Light micrograph of several degenerating midpachytene spermatocytes at differently advanced levels of degeneration. Tubule of developmental step VII-IX.2. ×880. **b** Electron micrograph of a midpachytene spermatocyte in the final state of degeneration. *Arrow* indicates remnants of prominent Golgi complex. ×4600. **c,d** Degenerating meiotic metaphase spermatocyte within a step XIII.2 tubule. Aligning of chromosomes at the metaphase plate as well as spindle fibers and one of the centrioles can still be seen despite the overall advanced degenerative state of the cell. **c** Light micrograph, ×880. **d** Electron micrograph, ×4600

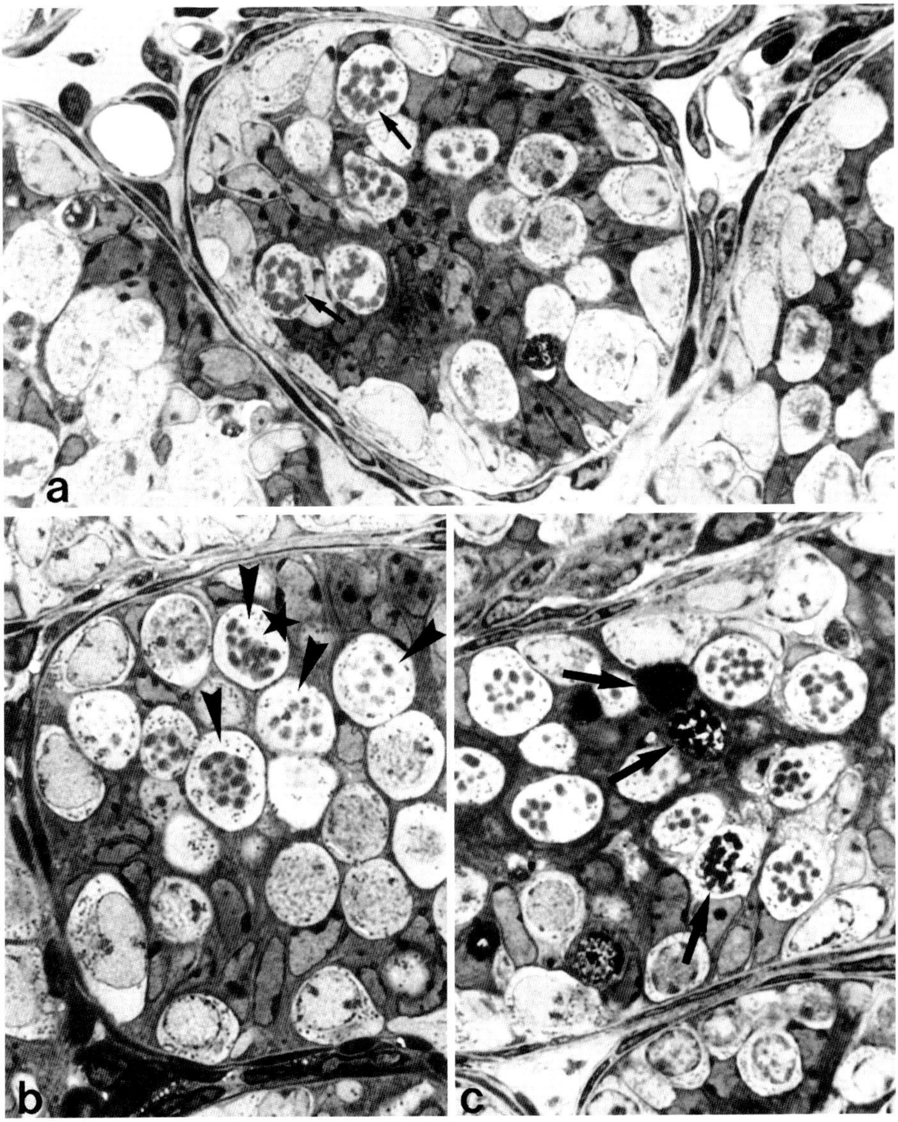

Fig. 24a–c. Light micrographs of arrested germ cell mitoses. Day 16. ×900. **a** Tubule of step IX-XII.1 (leptotene spermatocytes leading) containing several arrested mitoses. *Arrow* shows rosette-like formation of condensed chromosomes. **b** Tubule of step XII.1 (early-zygotene spermatocytes leading) with several such cells showing the process of increasing condensation of chromosomes and beginning dissolution of the nuclear membrane (*arrowheads*). *Asterisk* indicates cell depicted ultrastructurally in Fig. 26b. **c** Tubule of step IX-XII.1 (leptotene spermatocytes leading). Lying among several other arrested germ cell mitoses, three of these cells exhibit an advanced degenerative state (*arrows*)

Fig. 25a,b. Ultrastructure of arrested germ cell mitoses. **a** Characteristic, rosette-like arrangement of chromosomes and central position of all four centrioles. Some microtubules run radially towards the chromosomes. ×10,350. **b** Open intercellular bridges (*arrows*) between arrested mitoses, one of which is partly occupied by chromosomal material. ×5000

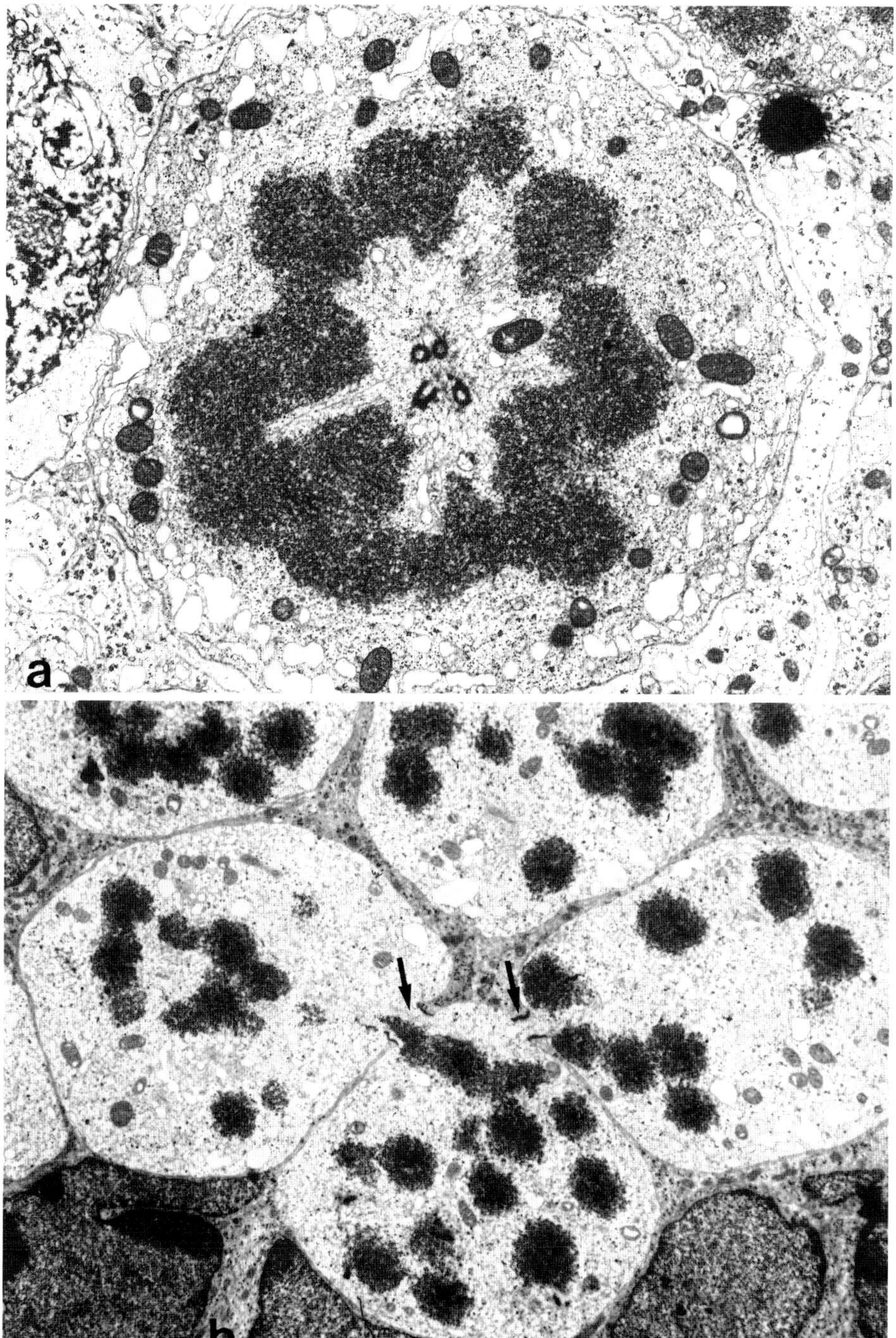
a
b

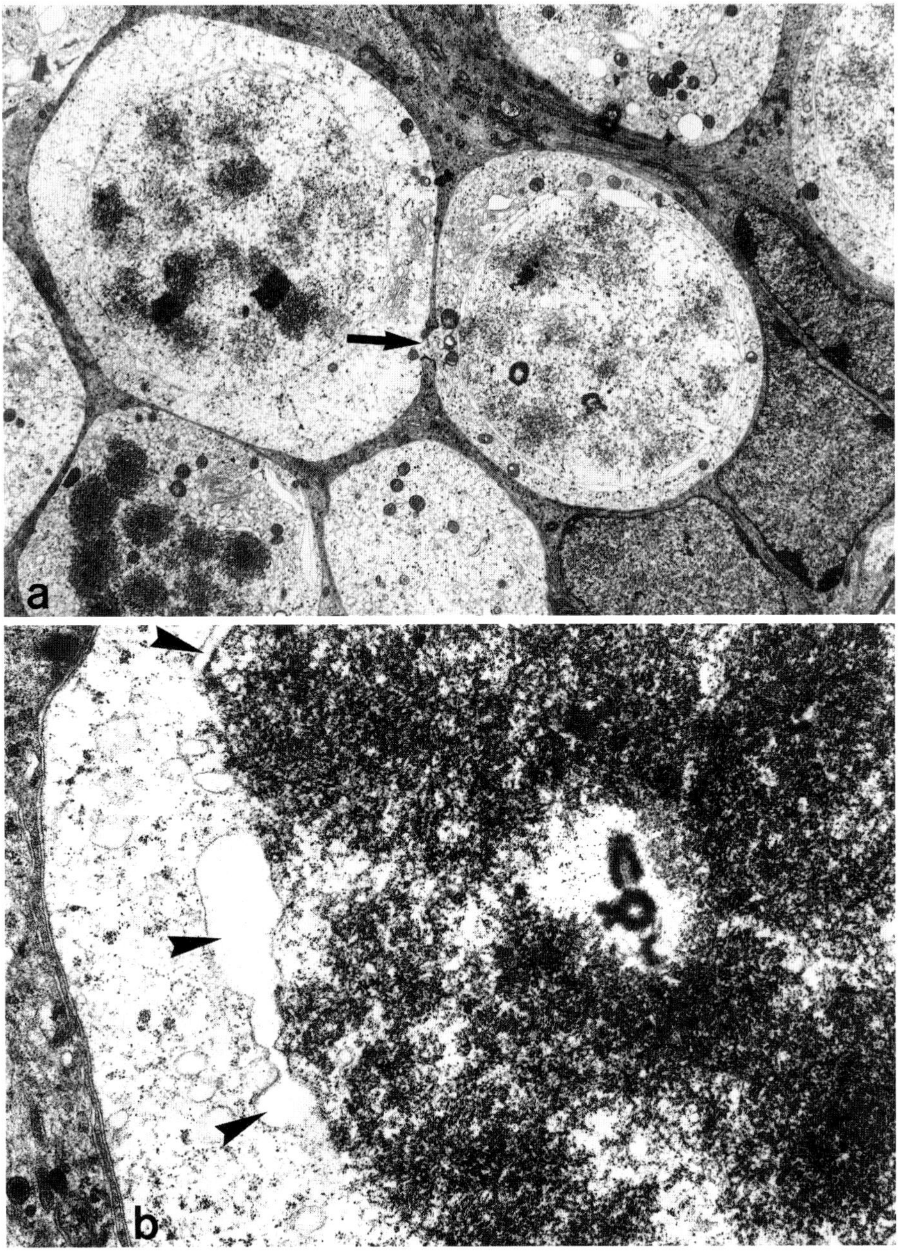

Fig. 26a,b. Ultrastructure of arrested germ cell mitoses. **a** Commencing, coarsely patched condensation of chromatin within a nucleus that is still completely enveloped. *Arrow* indicates intercellular bridge. ×3500. **b** Detail of the asterisk-labeled cell shown in Fig. 24b. Fragmentation of perinuclear cistern (*arrowheads*) indicates dissolution of nuclear membrane, while two of the centrioles have already attained a central position among the chromosomes. ×20,150

Table 3. Occurrence of arrested germ cell mitoses

Developmental step	Leading spermatocytes	Age (dpp) 14	15	16	17	18	19	20
VII-IX.1	Preleptotene	3	3	4	3	3	–	–
	Preleptotene/ leptotene	8	7	9	6	11	–	–
IX-XII.1	Leptotene	11	7	10	11	7	9	–
XII-XIII.1	Zygotene	3	8	2	5	4	16	25

For each of the days post partum (*dpp*) 14–20, 25 tubule sections containing this cell form were analyzed with respect to the associated spermatocyte types and were each assigned to one of the developmental steps of the seminiferous epithelium.

the preleptotene to the leptotene phase or have reached the fully developed leptotene phase, respectively.

3.6.3.4 Contact to Lanthanum

In sections of testicular tissue processed with a fixative containing lanthanum, the tracer has free access to the intercellular spaces surrounding arrested germ cell mitoses and adjacent spermatocytes (Fig. 27), indicating that the Sertoli cell barrier is not yet (completely) established in the tubule segments involved.

3.6.4 Spermatids

Most of the frequently observed instances of spermatid degeneration appear in connection with the temporary developmental disturbance of spermatids of step 6 or 6/7 (see Sect. 3.3). Round spermatids are affected up to day 30, while elongating ones are involved on the days that follow.

The prevailing degenerational pattern of spermatids prior to the commencement of elongation is the appearance of bi- or multinucleated cells exhibiting various nuclear or acrosomal deformations (Fig. 28). In single tubule segments, these cells may accumulate and form giant cells (Fig. 28a). Most cells can be assigned to step 6 or 6/7 of spermiogenesis, but younger spermatids may also occasionally be affected (Fig. 28b). Acrosome vesicles frequently fuse and thus tightly connect the matching nuclei (Fig. 28b,c). Even when strongly protruding into the tubular lumen, bi- or multinucleated spermatids are usually completely surrounded by slender lamellae of Sertoli cell cytoplasm (Fig. 28b,c) and thus remain integrated within the epithelium.

Elongating spermatids, appearing from day 32 onwards, sometimes also show degenerative alterations. These are various nuclear or acrosomal malformations in spermatids of steps 7–12 (Fig. 29a) and occasionally bi- or multinucleation (Fig. 29b,c).

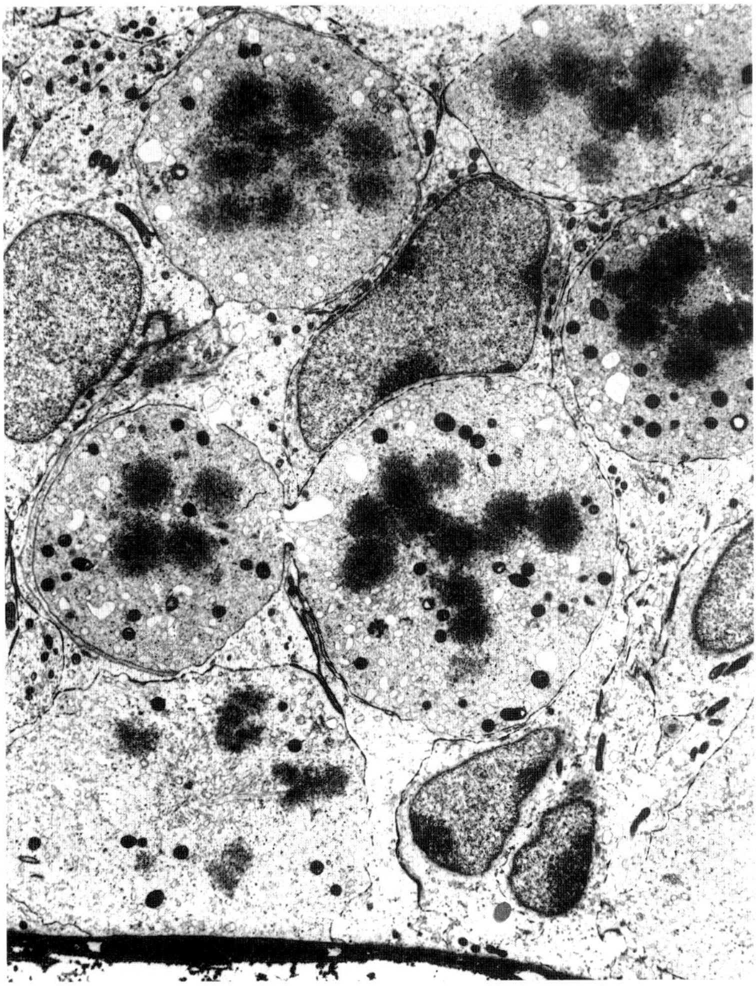

Fig. 27. Free passage of lanthanum throughout the intercellular spaces around arrested mitoses. ×4700

Such cells may be prematurely released from the epithelium, then lying within the tubular lumen.

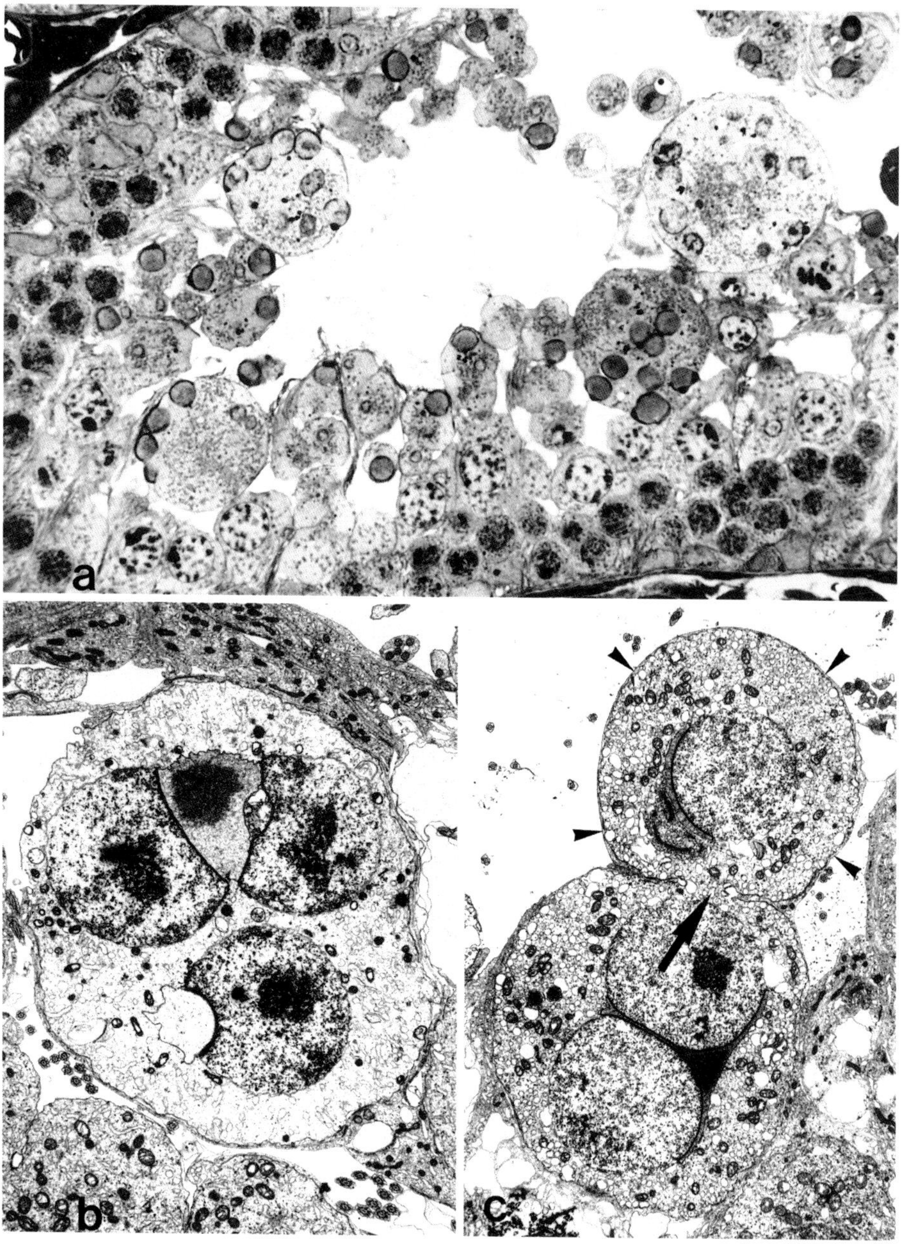

Fig. 28a–c. Degeneration of round spermatids. **a** Light micrograph of a tubule containing multiple multinucleated spermatid giant cells. Associated germ cell types conform to stage XIII. ×570. **b** Ultrastructure of a degenerating, three-nucleated spermatid of step 3–4 of spermiogenesis. ×3500. **c** Two bridge-connected (*arrow*) spermatids of step 6. Both nuclei of the cell below are tightly fixed to one another by a large acrosome vesicle. A slender lamella of a Sertoli cell completely encompasses the upper cell (*arrowheads*), protruding into the tubular lumen. ×2700

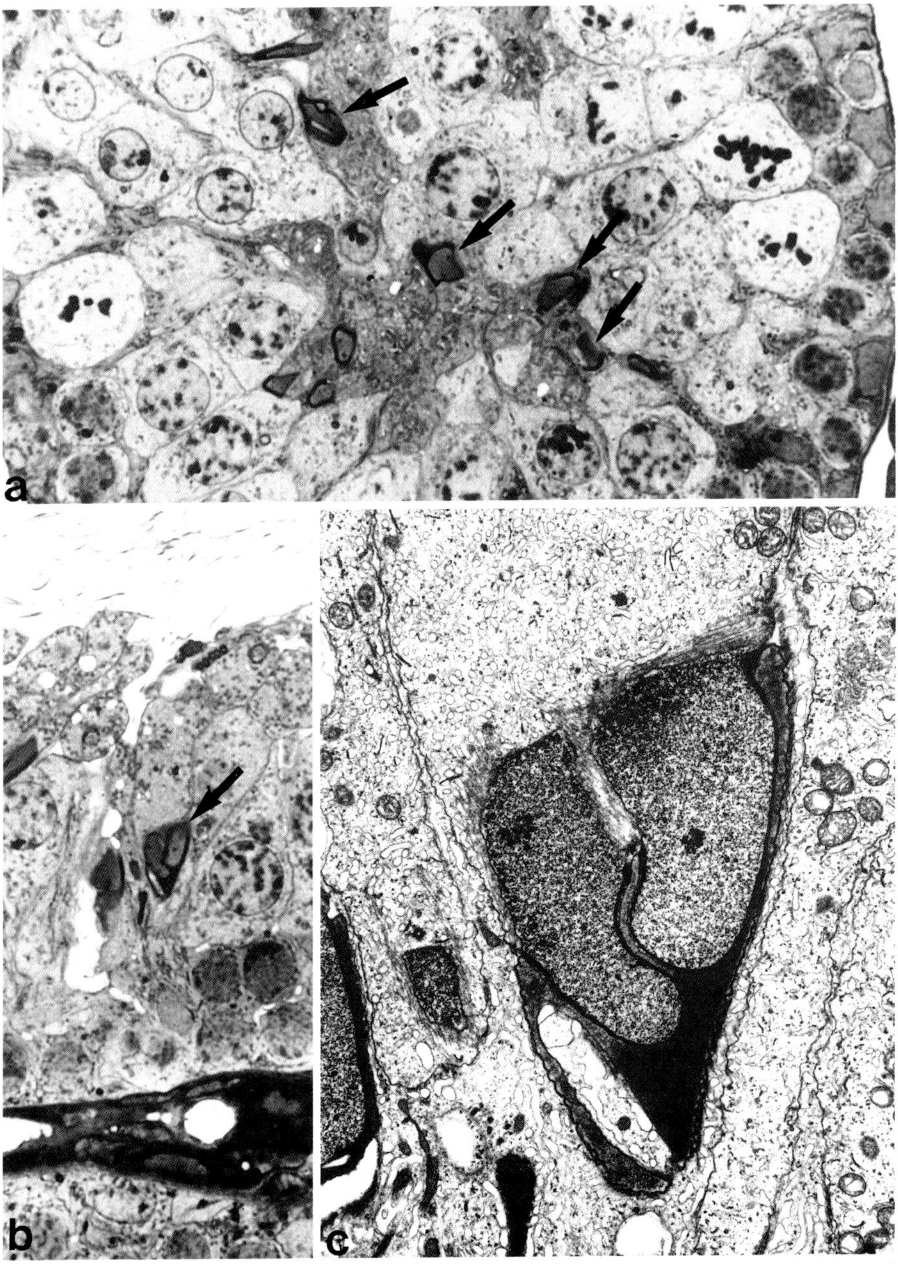

Fig. 29a–c. Degeneration of elongating spermatids. **a** Tubule section containing several deformed spermatids (*arrows*) derived from cells of spermiogenic steps 9–11. The developmental level of the associated germ cells conforms to stage XIII. Day 36. ×910. **b,c** Binucleated spermatid of step 9–10 (*arrow* in **b**), showing nuclear and acrosomal deformation and fusion of acrosome vesicles. Day 34. **b** Light micrograph, ×800. **c** Electron micrograph, ×6200

3.7
Clonal Mode of Germ Cell Proliferation

Pubertal germ cells form cell clones by means of intercellular bridges (Fig. 30a,b; see also Fig. 20a) evolving from incomplete mitotic and meiotic cytokineses. In the course of subsequent cell divisions of the cells involved, existing bridges persist but are temporarily closed by bridge-partitioning complexes (Weber and Russell 1987; Miething 1995; Fig. 31b). Intercellular bridges even persist for a certain period when the germ cells joined have entered a degenerative process (Fig. 31c,d; see also Fig. 28c). Irrespective of the type of germ cell involved, degenerating cells connected by intercellular bridges usually exhibit exactly the same mode and stage of degeneration (Figs. 28c, 31d).

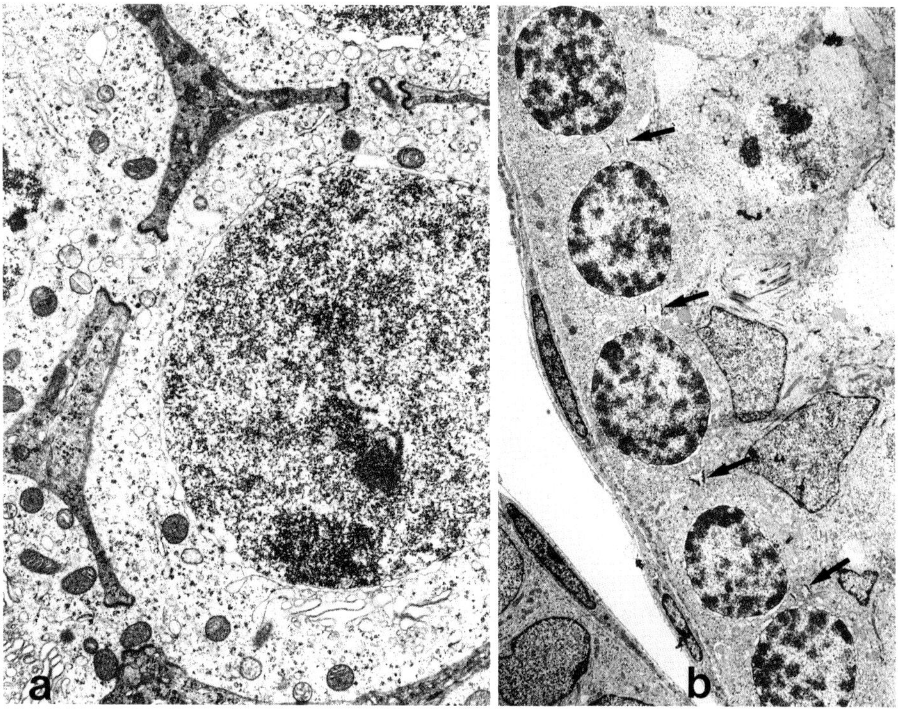

Fig. 30a,b. Intercellular bridges. **a** Primary spermatocyte being connected to neighboring cells (of the same clone) by three, freely passable intercellular bridges. ×7400. **b** Five bridge-connected (*arrows*) B-spermatogonia which synchronously exhibit prophase of mitotic division. ×2150

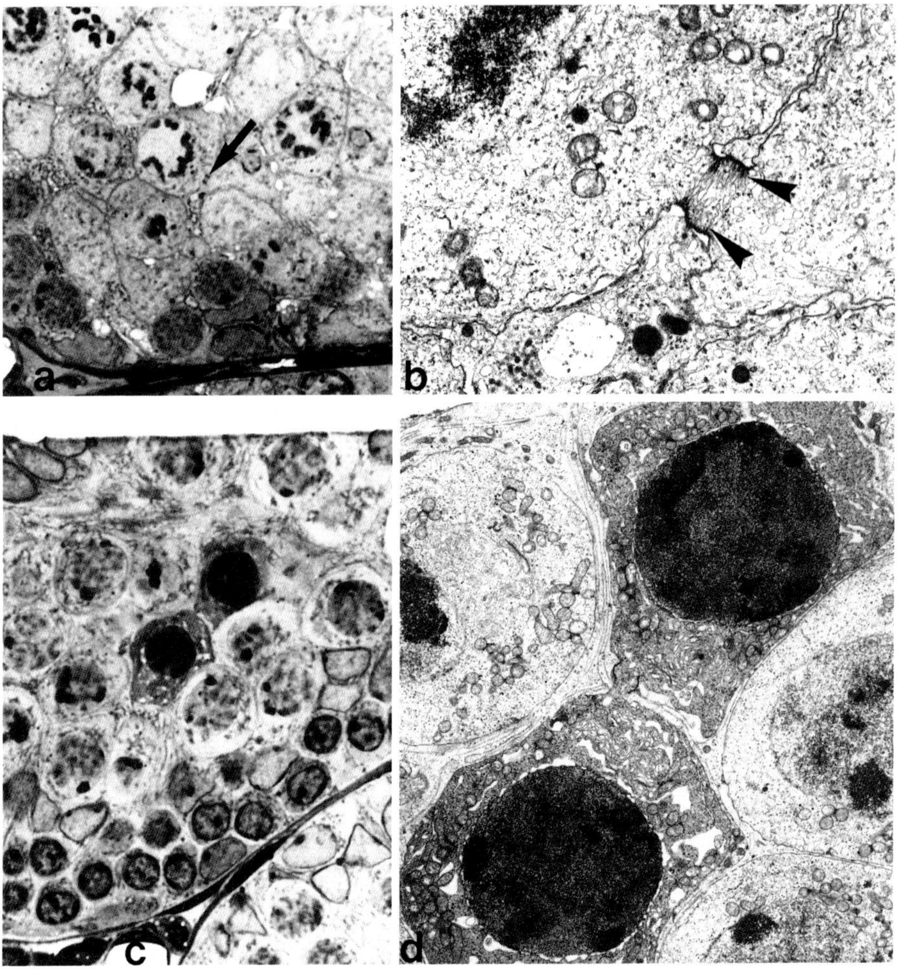

Fig. 31a–d. Intercellular bridges. **a,b** Bridge joining two diplotene/prometaphase spermatocytes (*arrow* in **a**). *Arrowheads* in **b** indicate emerging bridge-partitioning complex. **a** Light micrograph, ×710. **b** Electron micrograph, ×7000. **c,d** Degenerating pachytene spermatocytes connected by a largely intact intercellular bridge. **c** Light micrograph, ×900. **d** Electron micrograph, ×3450

3.8 Proliferative Activity of the Adult Seminiferous Epithelium

Proliferative data concerning the adult seminiferous epithelium were reviewed as a background for the assessment of pubertal germ cell kinetics. Table 4 summarizes the determined S-phase indices according to germ cell types and to spermatogenic stages. A-Spermatogonia show medium to high labeling indices in stages XII–II and stages VII–IX, with the highest value of about 31% in stage I. In-Spermatogonia show increasing indices during stages II–IV up to nearly 18% in stage IV. B-Spermatogonia actively proliferate in both the stages in which they are present, generally exhibiting

Table 4. S-phase indices of germ cells of the adult seminiferous epithelium with respect to germ cell types and to spermatogenic stages

Stage	Leading germ cells	S-phase labeling indices (%)			
		A-Spermatogonia	In-Spermatogonia	B-Spermatogonia	Preleptotene spermatocytes
I	spt-1,14	31.0±20.3	–	–	–
II	spt-2,14	9.9±16.5	1.8±6.0	–	–
III	spt-3,15	0	13.7±20.9	–	–
IV	spt-4,15	0	17.9±17.9	–	–
V	spt-5,16	0	–	41.0±23.1	–
VI	spt-6,16	4.1±10.3	–	24.5±22.1	–
VII	spt-7,17	6.0±12.5	–	–	14.1±15.7
VIII	spt-8	12.2±17.4	–	–	13.1±17.4
IX	spt-9	7.9±13.2	–	–	14.5±15.8
X	spt-10	3.6±5.7	–	–	–
XI	spt-11	0	–	–	–
XII	spt-12	10.9±16.6	–	–	–
XIII	spt-13	16.3±21.4	–	–	–

Indices are given as mean values±SEM.

the highest labeling index (41%) in stage V. Postmitotic/preleptotene spermatocytes are labeled at a rate of about 14% in all the three stages in which they appear (stages VII, VIII, and IX). In general, labeling of germ cells is highly heterogeneous among different tubule sections belonging to one and the same spermatogenic stage. Correspondingly, unlike the zero value in Table 4, all of the labeling values are characterized by a fairly high SEM value.

Without exception, adult Sertoli cells do not exhibit labeling.

3.9 Proliferative Activity of the Pubertal Seminiferous Epithelium

3.9.1 Germ Cells

Proliferation of pubertal germ cells was analyzed in close relationship to the morphologically defined developmental steps of the seminiferous epithelium (see Sect. 3.2). A survey of the S-phase indices calculated for the different germ cell types within the subsequent developmental steps is given in Table 5. For each of the steps constituting

Table 5. S-phase indices of germ cells determined in the pubertal seminiferous epithelium with respect to germ cell types and to developmental steps of the seminiferous epithelium

Developmental step	Leading germ cells	S-phase labeling indices (%) A-Spermatogonia	In-Spermatogonia	B-Spermatogonia	Preleptotene spermatocytes
I.1	A-spg	36.3±15.5	-	-	-
I-IV.1	In-spg	-	19.7±18.4	-	-
V-VI.1	B-spg	-	-	50.9±28.0	-
VII-IX.1	PL-spc	-	-	-	39.9±19.9
IX-XII.1	L-spc	-	-	-	-
XII-XIII.1	Z-spc	-	-	-	-
I.2	A-spg/P-spc	29.5±24.6	-	-	-
I-IV.2	In-spg/P-spc	-	28.2±15.3	-	-
V-VI.2	B-spg/P-spc	-	-	46.8±33.4	-
VII-IX.2	PL-spc/P-spc	-	-	-	27.9±33.8
IX-XII.2	L-spc/P-spc	-	-	-	-
XII-XIII.2	Z-spc/P-spc	-	-	-	-
XIII.2	Di-spc/SII	-	-	-	-
I.3	spt-1	34.0±25.7	-	-	-
II.3	spt-2	15.2±15.6	9.3±12.6	-	-
III.3	spt-3	3.5±9.8	16.7±22.4	-	-
IV.3	spt-4	2.4±6.6	20.0±18.7	-	-
V.3	spt-5	0	-	43.2±27.2	-
VI.3	spt-6	11.8±17.4	-	41.1±23.8	-
VII.3	spt-7	10.2±24.5	-	-	19.9±17.9
VIII.3	spt-8	8.7±19.2	-	-	17.5±19.5
IX.3	spt-9	12.5±16.5	-	-	11.4±12.6
X.3	spt-10	3.0±13.2	-	-	-
XI.3	spt-11	4.4±12.1	-	-	-
XII.3	spt-12	13.6±16.6	-	-	-
XIII.3	spt-13	12.9±15.9	-	-	-
I.4	spt-1,14	25.4±17.3	-	-	-
II.4	spt-2,14	16.2±26.3	8.5±10.3	-	-
III.4	spt-3,15	0	17.1±18.0	-	-
IV.4	spt-4,15	0	16.8±17.9	-	-
V.4	spt-5,16	0	-	31.0±20.1	-
VI.4	spt-6,16	1.2±2.6	-	38.6±18.5	-
VII.4	spt-7,17	5.5±8.9	-	-	18.7±14.6

Indices are given as mean values±SEM.

the first and the second cycle of the seminiferous epithelium (steps I.1 to XII-XIII.1 and steps I.2 up to XIII.2), the leading germ cells are considered. In the subsequent steps, all types of germ cells involved in proliferation are evaluated separately. Tubule segments showing any of the non-stage-conforming germ cell associations characterized in Sect. 3.3.3 are not included in the quantitative evaluation.

During the first two cycles (steps I.1–XIII.2), the leading A-, In-, and B-spermatogonia and postmitotic/preleptotene spermatocytes, respectively, show high labeling rates which usually exceed the corresponding values found in adult stages. From the third cycle onwards (beginning with step I.3), most labeling indices are comparable to those in the corresponding adult stages. However, A-spermatogonia also show a moderate labeling activity in steps III.3, IV.3, and XI.3. As in the adult seminiferous epithelium, labeling indices generally show high SEM values.

The pairs of corresponding micrographs seen in Fig. 32 each show S-phase labeling and morphology of germ cells in several developmental steps of the seminiferous epithelium, i.e., stepsV-VI.1 (Fig. 32a,b), I.2 (Fig. 32c,d), VII-IX.2 (Fig. 32e,f), and II.3 (Fig. 32g,h). With regard to the labeling pattern within nuclei, the staining may either be restricted to the nuclear periphery (Fig. 32a,g) or, in other cases, may consist of more or less coarse granules distributed evenly over the nucleus (Figs. 32c,e). Apoptotic cells may also occasionally be labeled.

3.9.2 Sertoli Cells

Proliferative activity of Sertoli cells is confined to the period up to day 16 and is only found in tubule segments in which the leading germ cells have not yet entered the leptotene phase of meiosis I. Table 6 shows the labeling indices of Sertoli cells accord-

Table 6. S-phase indices of Sertoli cells on days 12 to 17 related to developmental state of the associated leading germ cells and to age level

Age (dpp)	S-phase labeling indices (%) of Sertoli cells[a]				
	A-spermatogonia (step I.1)	In-spermatogonia (step I-IV.1)	B-spermatogonia (step V-VI.1)	Preleptotene spermatocytes (step VII-IX.1)	Leptotene spermatocytes (step IX-XII.1)
12	7.9±2.1	–	–	–	–
13	5.8±2.2	6.3±3.3	–	–	–
14	5.1±0.9	5.4±1.2	4.9±1.2	–	–
15	–	2.6±2.2	2.8±1.9	2.0±2.2	0
16	–	–	1.9±1.2	0.4±1.0	0
17	–	–	–	0	0

Indices are given as mean values±SEM.
dpp, days post partum.
[a]In tubular sections with the given spermatogonia and spermatocytes as the leading germ cells.

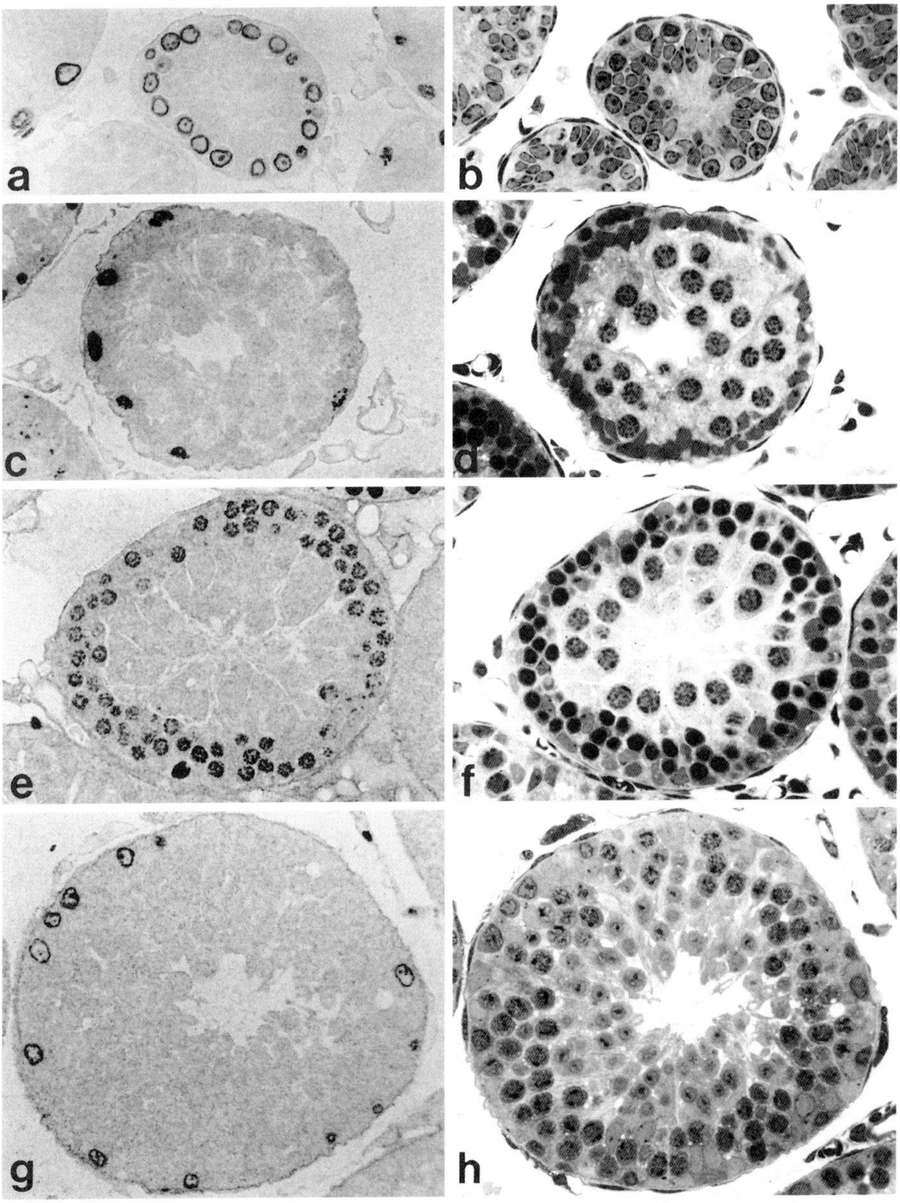

Fig. 32a–h. Proliferative activity of pubertal germ cells. Pairs of corresponding micrographs illustrate S-phase labeling (**a,c,e,g**) and morphology (**b,d,f,h**) of germ cells, respectively. ×400. **a,b** Step V-VI.1, B-spermatogonia labeled. **c,d** Step I.2, A-spermatogonia labeled. **e,f** Step VII-IX.2, postmitotic/preleptotene spermatocytes labeled. **g,h** Step II.3, In-spermatogonia labeled

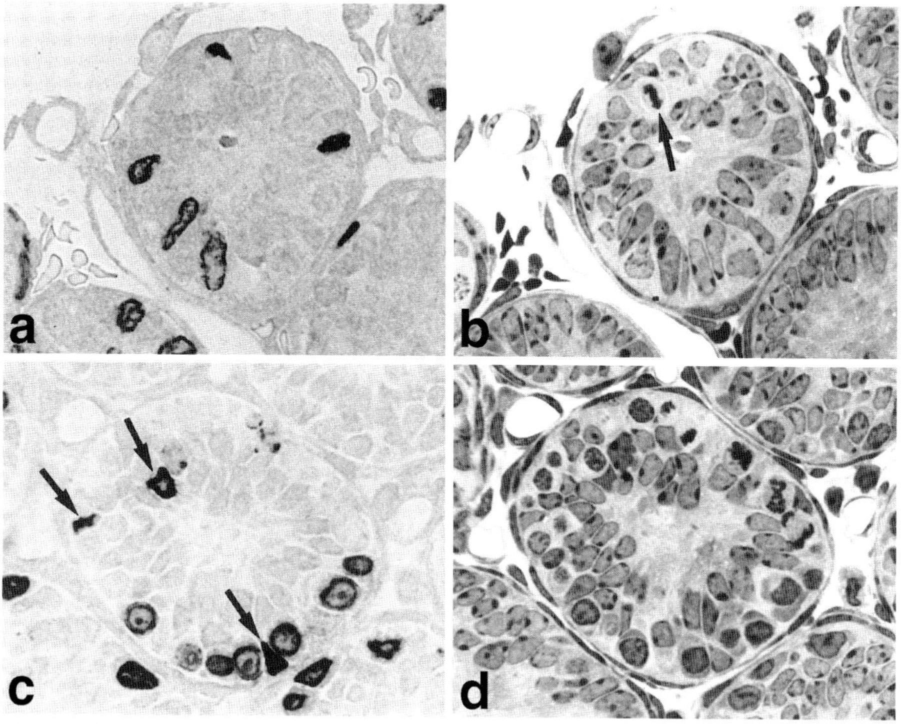

Fig. 33a–d. Proliferative activity of pubertal Sertoli cells. Pairs of micrographs document S-phase labeling (**a,c**) and morphology (**b,d**), respectively. ×480 **a,b** Step I.1 tubule containing several labeled Sertoli cells. *Arrow* in **b** indicates Sertoli cell mitosis. **c,d** Step I-IV.1 tubule showing three labeled Sertoli cells (*arrows*) among several labeled In-spermatogonia.

ing to age and developmental state of the locally associated germ cells. There is a gradual decline in labeling values, and from day 17 onwards no more Sertoli cells are labeled. On days 15 and 16, when the leading tubule segments have reached developmental steps IX-XII.1 onwards (see Fig. 12), appearance of S-phase-labeled Sertoli cells is confined to those tubule sections showing the steps up to and including VII-IX.1.

The corresponding micrographs presented in Fig. 33a–d show several labeled Sertoli cells in a tubule of developmental step I.1 (in addition to A-spermatogonia) and of step I-IV.1 (in addition to several labeled In-spermatogonia), respectively.

4 Discussion

4.1 Considerations on the Species and on Methodical Procedures

Regarding the well-ordered and relatively short developmental course of prespermatogenesis and the clear organizational pattern of adult spermatogenesis, the golden hamster (*Mesocricetus auratus*) largely resembles other rodent species, such as the rat and mouse. The seasonal cycle of the golden hamster's reproductive activity did not affect the present investigation because of the artificially controlled cycle of 14 h light and 10 h darkness. As the period of prespermatogenesis has been studied by the present author in the golden hamster (Miething 1989, 1990, 1992a, 1993), the choice of this species for the present investigation allows direct reference to the foregoing period of testicular development.

Among the methods employed (light and electron microscopy, immunohistochemistry, enzyme histochemistry), histological processing of individual (parts of) tubules and thin sectioning of reembedded semithin sections for electron microscopy will be discussed briefly. Flat plastic embedding and longitudinal sectioning of isolated tubules (Miething 1996) combines the method of aldehyde/osmium fixation and plastic embedding with the technique of sectioning individual tubules lengthwise (Perey et al. 1961). This latter method allowed the spatial organization of the adult rat seminiferous epithelium to be analyzed in detail ("wave of the seminiferous epithelium"; Perey et al. 1961). However, a more sophisticated histological method seemed necessary in order to be able to examine discrete morphological characteristics of the early germ cell types present in the pubertal testis, as the PAS–hematoxylin-based analysis of acrosome differentiation in spermatids is hardly applicable here due to the complete or partial absence of spermatids.

Thin sectioning of reembedded semithin sections (e.g., Schabtach and Parkening 1974; Maunsbach 1978; Larramendi 1985; Miething 1992b; DeFelipe and Fairen 1993) has repeatedly proven a valuable tool in morphological studies, even allowing targeted ultrastructural analysis of structures rarely found in sectioned material. In the present study, this technique was especially helpful for characterizing certain forms of degenerating germ cells and for tracing the formation of the Sertoli cell barrier.

4.2 Developmental Dynamics of the Seminiferous Epithelium

Pubertal maturation of the seminiferous epithelium is characterized by the concurrence and superimposition of two dynamic components:

1. The developmental progression of initial spermatogenesis at a given point of time has reached dissimilar levels in the different tubule sections within one and the same testis. Although this intratesticular variance (Sect. 3.2.4.2) has been mentioned in several studies on pubertal testicular development in the rat (Clermont and Perey 1957a; Russell et al. 1987; Neumann et al. 1993), it is still far from being comprehensively understood. This phenomenon probably reflects the cyclic dynamics of spermatogenesis already present while it is being established.
2. In the course of the period studied, a continuous maturational process occurs by which for the first time the germ cells pass through the complete developmental sequence of spermatogenesis up to initial spermiation of mature spermatids.

A key methodical feature of the present study was the investigation of the morphological development of the epithelium, and of the germ cells in particular, initially entirely independently of the temporal dimension. This facilitates analysis of the maturation of the seminiferous epithelium and at the same time allows the two developmental components described above to be distinguished. As a result of this procedure, a sequence of defined developmental steps of the seminiferous epithelium is established in strict accordance with the developmental level of the respective leading germ cells. Subsequently, the morphological data were projected onto the temporal course of the period studied and were thus assigned to concrete time points or phases within that period.

The other developmental parameters of the seminiferous epithelium studied, such as proliferative activity of germ cells and Sertoli cells, cellular degeneration, or formation of the Sertoli cell barrier, were also primarily related to the developmental state of the involved or directly neighboring germ cells and to the developmental step achieved by the local area of the seminiferous epithelium. Only in a second step were several of these parameters classified with respect to the temporal dimension of testicular maturation.

4.3 Germ Cell Development

4.3.1 Relation to Adult Spermatogenesis

Adult spermatogenesis of the golden hamster has been studied in detail both with regard to germ cell morphology and to the organizational pattern of germ cell development (Clermont 1954; de Rooij 1968; Clermont and Trott 1969; Schubert 1977). In the present investigation, this data is taken as a starting point from which to examine the initial establishment of spermatogenesis.

While the morphology and developmental succession of adult germ cells are readily comparable with the pubertal development, care should be taken in choosing which concept of cyclic organization of spermatogenesis to take as a basis for analysis. In order to interpret the present results, the concept presented by Clermont (1954) was chosen (see Sect. 1.2; Fig. 2), according to which the spermatogenic cycle is subdivided into 13 stages, largely based on the acrosomal and nuclear morphogenesis of the spermatids. This highly differentiated classification offers the advantage that the progress of pubertal germ cell maturation can be closely followed and even discrete developmental alterations can be recognized. This would only be possible to a lesser extent if other concepts of classification of the spermatogenic cycle were employed, such as those introduced by Roosen-Runge and Giesel (1950; rat, eight phases on the basis of spermatid maturation), Huckins (1978; rat, six stages on the basis of spermatogonial differentiation), or Bergmann (1987; Djungarian hamster, eight stages on the basis of spermatid maturation). When investigating the early developmental period of the seminiferous epithelium prior to the appearance of spermatids, however, even reference to the concept presented by Clermont (1954) requires further consideration of the present spermatocytes and/or spermatogonia. The definition of developmental steps in this context therefore only allows assignment to adult stage groups (see developmental steps I-IV.1 to XII-XIII.2,Table 1).

4.3.2 Pattern of Synchronization

The interplay of the diverse germ cell generations present within the pubertal seminiferous epithelium generally conforms to the synchronizational pattern of adult spermatogenesis. This applies both to the germ cells associated in a given localized area of the epithelium (vertical synchronization) and to the germ cell associations succeeding one another in the longitudinal direction of the tubule (horizontal synchronization).

4.3.2.1 Vertical Synchronization

From the beginning of pubertal spermatogenesis, neighboring germ cells develop in close coordination. This initially concerns the synchronous developmental progression of the cells belonging to one germ cell generation. The persisting intercellular bridges, which are freely passable for most of the cell cycle (Fawcett et al. 1959; Fawcett 1961; Weber and Russell 1987), play an essential role in synchronizing germ cell development within a clone (Fawcett et al. 1959; Gondos 1973, 1984). This role has been confirmed by the biochemical demonstration of mRNA transfer among the cells of a clone (Braun et al. 1989) and further substantiated by comparing cell counts of whole cell clones with the numbers of spermatids actually occupying (sub)segments of the adult rat seminiferous epithelium (Ren and Russell 1991).

In addition to this intraclonal tuning, the pattern of coordination of germ cell generations associated in a given area of the epithelium largely corresponds to the situation in the adult seminiferous epithelium. Although germ cell associations appearing in the course of the pubertal period are incomplete compared with the cellular composition of adult stages, as in most cases one or several types of germ cells are still missing, the germ cell associations can usually be assigned to one of the adult stages and are thus correctly said to exhibit stage conformity.

This relationship illustrates the high level of synchronization of germ cell development that is already present during this period, a feature which has been insufficiently analyzed in the previous literature. In their study on the immature rat testis, Clermont and Perey (1957a) also found indications of the presence of a cyclic organization in the initial course of germ cell development and basically related pubertal spermatogenesis to the organizational pattern of the adult. However, as the period studied was only divided into a fairly small number of age-groups and the investigation was restricted to paraffin histology, these results seem to be in need of reinterpretation to some extent. Between postnatal days 15 and 18, for instance, that study describes four different types of tubules containing the following germ cells: type 1 (A- and In-spermatogonia/pachytene spermatocytes), type 2 (A- and B-spermatogonia/pachytene spermatocytes), type 3 (A-spermatogonia/resting preleptotene and, in parts, pachytene spermatocytes), and type 4 (A-spermatogonia/leptotene and zygotene spermatocytes). The tubule types are thought to represent four successive stages of a cycle of the seminiferous epithelium. By analyzing these results in detail and applying the terminology introduced in the present study (for the golden hamster), the following classification of the types of tubules observed by Clermont and Perey (1957a) can be made: type 1 corresponds to developmental step I-IV.2, type 2 to step V-VI.2, type 3 to step VII-IX.1 (without pachytene spermatocytes) and to step VII-IX.2 (along with pachytene spermatocytes), respectively, and type 4 to steps IX-XII.1/XII-XIII.1. The chronological succession of tubule types should therefore be as follows: type 3 (without pachytene spermatocytes), type 4, type 1, type 2, and type 3 (with pachytene spermatocytes).

In agreement with the present results on the golden hamster, Clermont and Perey (1957a) determined the appearance of germ cell associations exhibiting stage conformity for the advanced pubertal period between days 33 and 45 in the rat. In addition, a quantitative distribution of associations comparable to the adult values was also evaluated for that period.

Other authors have followed the pubertal germ cell development in the rat and specified the most advanced types of germ cells present at different developmental times, but without further consideration of the pattern of synchronization of all germ cells present at a given epithelial site (Knorr et al. 1970; Russell et al. 1987). In a further study on pubertal germ cell maturation in the rat between days 25 and 41, Neumann et al. (1993) described eight transient germ cell associations (types A–H). Although they employed semithin sectioning histology, their morphological results remain vague and only partly elucidate the developmental behavior of germ cells in terms of stage conformity as shown in the present study. Only the last three pubertal cell associations (types F–H) are directly related to adult spermatogenic stages by Neumann et al. (1993).

4.3.2.2 Horizontal Synchronization

The spatial arrangement of pubertal germ cell associations along the length of the tubules, as perceivable from examination of individually flat-embedded tubule sections (Figs. 14, 15), already corresponds to the organizational pattern of the adult wave of the seminiferous epithelium (Perey et al. 1961). The cyclic pattern of germ cell development, demonstrated by Clermont and Perey (1957a) and by the present results, is thus also reflected in the topographical arrangement of the developmental steps of the epithelium. Adjacent tubule segments usually represent immediately successive developmental steps of the seminiferous epithelium. However, segmental boundaries are not as sharply demarcated as in the adult. This is presumably due to the missing spermatid generations or, in advanced developmental steps, to relatively low numbers of spermatids. Irrespective of the age of an animal and of the concrete succession of steps present, the developmental interval covered by the steps of a given tubule regularly amounts to one cycle length.

Testicular tubules not only undergo a considerable increase in diameter during pubertal development (Kluin et al. 1984; Vergouwen et al. 1991; see also Figs. 8–10, 32), but also grow to several times their original length (Clermont and Huckins 1961). The length of tubule segments occupied by one developmental step and the length of a whole spermatogenic wave therefore probably increase in the course of the period studied, finally reaching the adult values. In the rat, the average length of one wave of the seminiferous epithelium amounts to 2.6 cm (Perey et al. 1961), while in the golden hamster this value has not yet been determined.

4.3.3 Variance

The existence of different developmental levels of the seminiferous epithelium in tubule sections of one and the same pubertal testis has already been observed in the rat (Clermont and Perey 1957a; Neumann et al. 1993) and the cat (Sanchez et al. 1993). The most detailed data in this connection was provided by Neumann et al. (1993), who defined eight transient associations of germ cells. They appear successively between days 25 and 41, are manifest for several days each, and are lost in the same order of succession. This developmental behavior of the seminiferous epithelium is confirmed and further specified by the present results. The interval covered by the developmental differences among the tubule sections within a given testis, at first imposing as an intratesticular variance, generally amounts to the length of one spermatogenic cycle (see Figs. 12, 13). However, as almost exactly the same developmental interval is already present within each of the tubules (see Sect. 4.3.2.2), this intra*testicular* variance actually reflects the succession of developmental steps within each single tubule, i.e., an intra*tubular* variance (or wave of the seminiferous epithelium). The tubules of a given testis are thus equal with respect to the succession of developmental steps present; on the whole, they develop synchronously within one testis.

4.4
Germ Cell Proliferation

The proliferative activity of germ cells was studied by determining the S-phase indices of A-, In-, and B-spermatogonia and postmitotic/preleptotene spermatocytes (see Tables 4, 5). This allowed the proliferative behavior of these germ cells during the pubertal and the adult spermatogenic cycle to be compared; a further analysis of the proliferative pattern of the generations of undifferentiated A-spermatogonia (stem cells) or a detailed characterization of the premeiotic S phase of spermatocytes was not intended on the basis of this data.

The present S-phase values determined for adult germ cells (Table 4) correspond closely to the germ cell mitotic indices counted by Clermont (see Table II in Clermont 1954) in the single spermatogenic stages of the golden hamster. However, S phases partly precede the respective mitotic peaks by several stages and show a broader distributional scatter. Clermont (1954) thus observed divisions of A-spermatogonia in stages IX and X as well as in stage XIII, but S phases of these cell types appear as early as stage VI and stage XII onwards, respectively. On the one hand, this might indicate a certain flexibility in the temporal succession of S phase and subsequent mitosis (duration of G_2 phase). On the other hand, this feature might possibly be due to a less strict stage dependency of spermatogonial proliferation, as has recently been described for bovine spermatogenesis (Wrobel et al. 1995, 1996).

Comparison of S-phase indices of pubertal (see Table 5) and adult (Table 4) germ cells reveals a high level of agreement. As developmental steps during the two initial cycles (steps I.1 to XII-XIII.1 and steps I.2 to XIII.2) generally do not correspond to single stages but to groups of stages, the data concerned is not directly comparable to the adult data. Only the S-phase indices of the leading germ cells were determined for this period. Nevertheless, it is possible to detect conformity with the stage-bound S-phase values of adult germ cells.

From the third cycle onwards (beginning with step I.3), evaluation was carried out in an exact stage-related manner and consistently yielded S-phase values in the range of adult results. A-Spermatogonia show no S-phase activity in only one step of the third cycle (step V.3), while in the adult S-phase activity is absent in several stages (III, IV, V, and XI). Cautiously interpreting the values (characterized by high SEM levels), this might be indicative of a somewhat lower degree of coordination of A-spermatogonial proliferation during this period.

Almost from their first appearance during the spermatogenic cycle onwards (adult: from stage VII; pubertal: from step VII.3 onwards), postmitotic/preleptotene spermatocytes exhibit labeling indices of between 11% and 20%. On account of their nuclear morphology, these cells have been subdivided into postmitotic (resting) spermatocytes (Clermont 1954; Hilscher and Makoski 1968; Schubert 1977) and the subsequent preleptotene spermatocytes proper. In both subphases, they are partly labeled. This is in agreement with a study presented by Wartenberg et al. (1998) on the immature rat ovary, according to which the premeiotic S phase of oocytes extends throughout the late postmitotic and early preleptotene period.

4.5 Delayed Spermatid Maturation and 11β-Hydroxysteroid Dehydrogenase Activity in Leydig Cells

Between days 26 and 30, those spermatids that have reached steps 6 or 6/7 of spermiogenesis (and representing the leading generation of germ cells) are selectively affected by a complete maturational arrest. This partly coincides with severe signs of degeneration of the cells involved (see below). This phenomenon is interesting in several respects: it allows the study of a unique form of physiologically occurring, partial desynchronization of germ cell generations; in addition, it points to the specific conditions and dependencies of the (commencing) process of spermatid elongation from step 7 of spermiogenesis onwards.

4.5.1 Desynchronization of Germ Cell Generations

Signs of desynchronization of the seminiferous epithelium comparable to the present results are mentioned in the literature only by Russell et al. (1987) in the pubertal rat. These authors observed arrested and partly malformed spermatids of steps 9–13 on day 32 within otherwise more advanced stages of spermatogenesis. However, due to a rather poorly differentiated time scale (days 20, 32, and 44), the formation and dynamics of this phenomenon were not addressed in their study. The progressive developmental discrepancy between stagnant spermatids and the other, normally developing germ cell generations is only detectable if a sufficiently close-meshed sequence of age levels is examined. This is especially true for the assessment of the many retarded and frequently degenerating elongating spermatids of steps 8–12. Judged on the basis of the present data, these cells had temporarily been arrested at step 6 (or 6/7) of spermiogenesis and entered a process of elongation beyond day 30 that was eventually abortive.

The normal further development of spermatogonia and spermatocytes between days 26 and 30 indicates that the younger germ cells, at least in this special constellation, are independent of the level of spermatid differentiation. Mutual control by the different generations of germ cells as a means of establishing and maintaining their relative synchronization does not seem to exist. This impression is supported by the stage-conforming establishment of the seminiferous epithelium long before complete germ cell associations appear (Clermont and Perey 1957a; present results). Furthermore, this view is in accordance with several experimental investigations which prove that the pattern of coordination of the remaining types of germ cells is not lost even when single germ cell generations are completely eliminated by radiation (Dym and Clermont 1970), heat (Chowdhury and Steinberger 1964), or administration of toxic substances (Bartlett et al. 1988; Russell et al. 1990). Thus, the invariable duration of the single phases of spermatogenesis in conjunction with the entry of new cell generations into the spermatogenic process at constant intervals presumably plays a key role in synchronizing germ cell development (Clermont 1972).

4.5.2
11β-Hydroxysteroid Dehydrogenase in Leydig Cells

The specific developmental blockage of spermatids is overcome at exactly the same time as 11β-HSD activity is established in the Leydig cells. When histochemically determining several nonspecific HSD in Leydig cells of the pubertal rat testis, Neumann et al. (1993) also found a conspicuous temporal concurrence between the first appearance of elongated spermatids and the maturation of 11β-HSD (around day 35 in that species). A developmental arrest of round spermatids on the preceding days cannot, however, be clearly inferred from the study by Neumann et al. (1993).

Two experimentally proven relationships suggest that these two coincident events (i.e., spermatid elongation and 11β-HSD maturation of Leydig cells) are causally related: (1) the onset of elongation of spermatids is highly testosterone dependent and (2) 11β-HSD represents an essential factor regulating the testosterone output of Leydig cells.

4.5.2.1
Testosterone Dependency of Spermatid Elongation

Transformation of round to elongated spermatids is blocked in hypophysectomized (Clermont and Morgentaler 1955; Bartlett et al. 1989) or LH-suppressed (Sun et al. 1990; O'Donnell et al. 1994) adult rats and is restored only after testosterone substitution. Testosterone levels of approximately 10%–20% (Cunningham and Huckins 1979b; Zirkin et al. 1989), 20%–40% (Awoniyi et al. 1989a,b), or 24%–46% (Sharpe et al. 1988) of the physiological value have been determined to be sufficient to maintain complete spermatogenesis, including spermatid transformation. The precise relationship between intratesticular testosterone levels and spermatogenesis is still controversial (Sharpe 1987; Rommerts 1988; Weinbauer and Nieschlag 1990). Furthermore, following the temporary selective destruction of rat pubertal Leydig cells by means of ethane dimethane sulfonate administration (Bartlett et al. 1986), a deficiency of spermatids of steps 8–9,i.e., at the onset of elongation, has been reported (Cameron et al. 1993). Finally, a direct influence of testosterone on the differentiation of elongating spermatids is indicated by the presence of androgen receptors within the nuclei of these cells (Wright and Frankel 1980; Vornberger et al. 1994; Haider et al. 1995).

4.5.2.2
Significance of 11β-Hydroxysteroid Dehydrogenase for Testosterone Production and Spermiogenesis

11β-HSD is able to oxidatively inactivate glucocorticoids (Monder and Shackleton 1984; Whorwood et al. 1992). Within the testis, it is present as type I 11β-HSD uniquely in Leydig cells (Phillips et al. 1989), where the predominance of its oxidative activity is physiologically maintained by active corticosterone (Gao et al. 1997). Leydig cells exhibit glucocorticoid receptors (golden hamster: Ortlip et al. 1981; rat: Stalker et al. 1989) and furthermore display a glucocorticoid-mediated suppression of androgen biosynthesis (Welsh et al. 1982). The physiological significance of Leydig cell 11β-HSD has been determined by means of hormone and enzyme quantifications in both stressed and unstressed rats (Monder et al. 1994a) and additionally following experimental administration of glucocorticoids (Monder et al. 1994b). The enzyme decreases intracellular glucocorticoid concentration and thus protects Leydig cells against the inhibitory effects of these hormones on testosterone secretion (Monder 1994a,b). 11β-HSD therefore plays an essential role in the regulation of the testosterone level, presumably also doing so during pubertal development. This is in agreement with measurements of blood testosterone levels in pubertal golden hamsters, which show a steep increase beyond day 30 (Vomachka and Greenwald 1979; Frungieri et al. 1996) and, in studies using less differentiated age steps, increasing values around day 28 (Miller et al. 1977; Clemens and Witcher 1985) or approximately from day 25 onwards (Urbanski et al. 1992).

Taken together, the above data support the view inferred from the present observations that a distinct increase in testosterone output is made possible only as a result of the maturation of Leydig cell 11β-HSD and, in turn, spermatids are induced to enter the previously blocked process of elongation. This relationship is schematically presented in Fig. 34. The "belated" Leydig cell maturation enforces a developmental delay of several days' duration on the leading germ cells and thus entails the demonstrated partial desynchronization of the seminiferous epithelium as well as a considerable germ cell loss; however, the biological significance of this delayed maturation remains unclear.

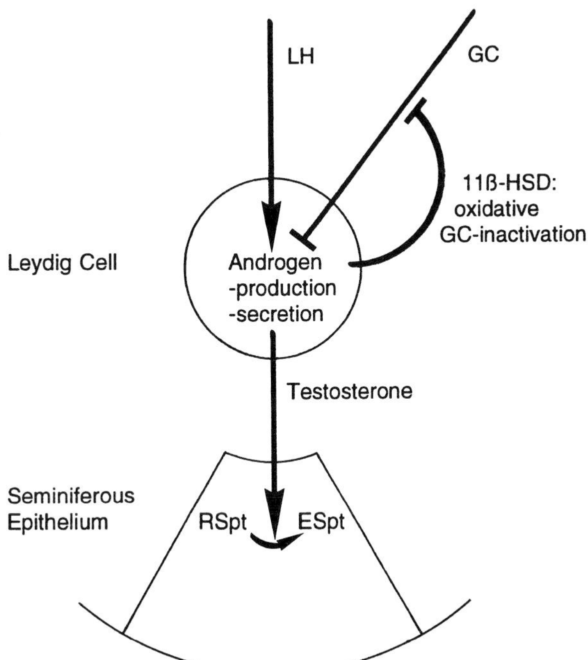

Fig. 34. Relationship between Leydig cell 11β-hydroxysteroid dehydrogenase (11β-HSD), testosterone output of Leydig cells, and elongation of spermatids as one parameter for progression of spermatogenesis. *LH*, luteinizing hormone; *GC*, glucocorticoids; *RSpt*, round spermatids; *ESpt*, elongating spermatids; *arrow*, stimulation; *bar*, inhibition

4.6 Degenerative Phenomena of the Pubertal Seminiferous Epithelium

With respect to both their morphological appearance and temporal distribution, the present observations of germ cell apoptoses and intrameiotically degenerating spermatocytes largely parallel the situation in the adult and are in agreement with previous results on pubertal germ cell degeneration (Allen and Altland 1952; Russell et al. 1987; Billig et al. 1995; Mori et al. 1997; Miething 1997). The following discussion of degenerative pubertal germ cells therefore focuses on the extensive loss of spermatids and on the arrest of germ cell mitoses, both of which are phenomena that seem to be specific to pubertal germ cell development.

4.6.1 Degeneration of Spermatids

The pronounced degenerative disposition of round spermatids (occurring up to day 30) and of elongating spermatids (occurring on the subsequent days) is closely related to the temporary maturational retardation of the cells involved between days 26 and 30. It is manifested by the frequent appearance of multinucleate and/or progressively degrading spermatids (see Figs. 28, 29). A corresponding accumulation of degenerating spermatids has also been reported in the pubertal seminiferous epithelium of the rat (Clermont and Perey 1957a; Russell et al. 1987) and the rabbit (Tsunenari and Kast 1992). This phenomenon seems to contribute to the marked reduction in the overall cellular output of the initial spermatogenic cycle and to the temporal difference between qualitative and quantitative completion of the establishment of spermatogenesis (Allen and Altland 1952; Robb et al. 1978; Zhengwei et al. 1990).

One possible reason for the increased loss of spermatids might be the direct relationship between testosterone deficiency and a reduced Sertoli cell-spermatid junctional interaction recently identified in vivo and in vitro (Cameron and Muffly 1991; Muffly et al. 1994). It is presumably caused by the dissociation of ectoplasmic specializations (Russell 1977b, 1993) under testosterone-suppressed conditions (Cameron et al. 1993). These adhesive structures are probably mediated by cell adhesion molecules such as N-cadherin (MacCalman et al. 1993; Newton et al. 1993) or $\alpha 6\beta 1$-integrin (Salanova et al. 1995) and usually form around the time that spermatid elongation begins.

Two possible explanations arise for the present observation that (nearly) all spermatids belatedly entering elongation after day 30 do nevertheless degenerate. First, these cells might have been irreversibly damaged during the phase of developmental retardation. The fact that many of the round spermatids degenerate before day 30 supports this assumption. Second, the non-stage-conforming local environment might interfere with the normal further development of the spermatids involved. Here, the Sertoli cell-controlled milieu probably plays a substantial role; this milieu is closely adapted to the progress of germ cell development in the sense of a Sertoli cell cycle and is thought to be particularly important for germ cells within the adluminal compartment of the seminiferous epithelium (see Parvinen 1993).

4.6.2
Arrested Germ Cell Mitoses

The arrested mitotic figures observed in the present study between days 14 and 20 in tubules of developmental steps VII-IX.1 to XII-XIII.1 have previously been described in the pubertal testis only with respect to the missing bridge-partitioning complex within the intercellular bridges connecting them (Miething 1995). Morphologically, they closely resemble the initial manifestation of a distinct degenerative pattern of mitotic oogonia in the immature ovary, the so-called atretic divisions (Beaumont and Mandl 1962). The latter cells were consistently found in the ovary of the rat (Beaumont and Mandl 1962; Franchi and Mandl 1963; Hilscher et al. 1974), golden hamster (Challoner 1974; Pützer 1986), and human (Baker 1963; Baker and Franchi 1967) at the time when the germ cells pass from mitotic proliferation towards onset of meiosis. They are always located next to leptotene, zygotene, or pachytene oocytes (Beaumont and Mandl 1962; Baker 1963; Hilscher et al. 1974). This closely corresponds to the temporal and spatial appearance of such cells observed in the present study, i.e., nearby spermatocytes which have initially reached the developmental level of entering meiosis.

Thus, irrespective of sex, this degenerative mode affects germ cells which enter a mitotic division while in immediate proximity to meiotic germ cells. According to the developmental pattern of oogenesis, this occurs only during a limited (prenatal) period in the ovary. In contrast, spermatogenesis is characterized by a continuous, close coexistence of mitoses (spermatogonia) and meioses (spermatocytes), which, however, are effectively separated by the Sertoli cell barrier in the adult seminiferous epithelium. It is only when spermatocytes initially enter meiosis during pubertal establishment of spermatogenesis that this barrier, as demonstrated, is not yet (completely) functional.

Taken together, these results point to the hypothesis that meiotic germ cells (or possibly the associated Sertoli cells) deliver a diffusible agent which, when it reaches neighboring germ cell mitoses, is able to specifically inhibit these mitoses. This might actually be based on the disturbance of the mitotic spindle apparatus, a supposition which is supported by the conspicuous similarity between this degenerative form and monopolar mitotic figures of colchicine-treated cells (see Chap. 10 in Dustin 1984).

Other features of the arrest of germ cell divisions suggest two ideas concerning Sertoli cell physiology. On the one hand, the Sertoli cell barrier might take on additional significance, i.e., shielding a (basal) mitotic milieu against an (adluminal) meiotic milieu within the seminiferous epithelium. As opposed to the functions hitherto ascribed to the barrier (Johnson 1970; Setchell 1980; Byers et al. 1993), the hindrance of a diffusion gradient directed from the adluminal towards the basal compartment is emphasized here.

On the other hand, our understanding of the fact that mitotic Sertoli cell proliferation definitively ceases immediately prior to the entry into meiosis of the neighboring spermatocytes could be improved. In the light of the hypothesis outlined above, the strictly postmitotic state of advanced pubertal and adult Sertoli cells seems to represent a biological necessity, since possible mitoses of these cells would be subjected to the obstructing influence on the part of adjacent spermatocyte meioses and would thus be irreversibly damaged. Further investigations will elucidate the exact origin of the disturbance in the mitotic spindle apparatus of the affected (male and female) germ cells and will have to prove the presumed influence of neighboring meioses or somatic cells.

4.7 Maturation of Sertoli Cells and Germ Cell Development

Sertoli cell maturation has been followed, in the present study, on the basis of two developmental parameters, i.e., establishment of the Sertoli cell barrier and proliferative behavior. Both the completion of the Sertoli cell barrier and the final waning of Sertoli cell propagation seem to be (at least temporally) related to initiation of germ cell meiosis.

4.7.1 Sertoli Cell Barrier

Morphological and functional maturation of the Sertoli cell barrier extends over a certain period of time in all mammalian species studied in this connection. Its establishment is completed between days 15 and 19 in the rat (Vitale et al. 1973; Bergmann and Dierichs 1983; Russell et al. 1989), between days 10 and 16 in the mouse (Nagano and Suzuki 1976), around day 20 in the golden hamster (Vignon et al. 1987), between days 15 and 21 in the guinea pig (Pelletier and Friend 1983), between week 10 and 11 in the rabbit (Sun and Gondos 1986; Viebahn and Miething 1991), and around week 20 in the dog (Connell 1980). This temporal pattern basically parallels the results obtained in the present study in the golden hamster, which also reveal that the barrier is established over a period of several days.

However, only a few authors (Connell 1980; Bergmann and Dierichs 1983) have reported a temporal succession of barrier formation according to the presently described heterogeneous progression of germ cell development in different tubule segments. In the dog, the Sertoli cell barrier is not established synchronously throughout the whole tubule; its formation is related to the neighboring spermatocytes initially reaching the zygotene phase (Connell 1980). In the rat testis, Bergmann and Dierichs (1983) observed a correlation between completion of the barrier and reaching of pachytene phase by local spermatocytes. This is in agreement with the present results ascertaining a partially effective barrier in tubule segments with leptotene and zygotene spermatocytes and a complete barrier in segments with pachytene spermatocytes; thus the pattern of establishment of the Sertoli cell barrier is always related to the developmental level reached by the spermatocytes in the initial spermatogenic cycle (Gondos and Berndtson 1993). The reason why this relationship was not taken into consideration in a number of former studies may be that the heterogeneous developmental progression of germ cells within different tubule segments has not been considered systematically.

The origin of the developmental connection described remains unclear. A conceivable simultaneous dependency of the two parameters, i.e., germ cell development and Sertoli cell barrier formation, on hormonal stimuli does not seem probable: Neither pubertal barrier formation (Vitale et al. 1973) nor barrier integrity in the adult epithelium (Hagenäs et al. 1978) are substantially impaired under gonadotropin- or androgen-deficient conditions. Furthermore, such a general regulation could not account for the different developmental levels (of germ cells and of barrier formation) of the tubule segments. A direct causal relation between local germ cell differentiation and barrier formation is also doubtful, since in pubertal rat seminiferous epithelium entirely lacking germ cells (Sertoli-cell-only testes) normal, albeit moderately delayed establishment of the barrier occurs (Vitale 1975; Fawcett et al. 1975, cited in Nagano and Suzuki 1976).

This experimental data, however, does not rule out a possible modulating influence of certain developmental steps of germ cells on the exact timing of barrier formation. Rather, a concept based upon such a differential modulation of Sertoli cell maturation is most likely suitable for reconciling the present results with the conditions of barrier formation cited above. This seems particularly conceivable as there are indications of modulating effects of certain germ cell types on other (secretory) functions of Sertoli cells; these are mediated by receptor-specific humoral (protein) factors (Le Maguer-esse and Jegou 1988; Onoda and Djakiew 1991) or by direct cell–cell contacts (Galdieri et al. 1984; Castellon et al. 1989).

4.7.2 Sertoli Cell Proliferation

Propagation of Sertoli cells is generally confined to the fetal and early postnatal period and terminates around the time of initial spermatocyte appearance. In the rat, this has been consistently established by counting Sertoli cell mitoses (Clermont and Perey 1957a) and by autoradiographic quantification of S phases (Steinberger and Steinberger 1971; Orth 1982). Comparable results have been determined in the mouse (Kluin et al. 1984; Vergouwen et al. 1991) and rabbit (Sun and Gondos 1981). The golden hamster also shows a steady decline in the S-phase indices of the Sertoli cells as early as during prespermatogenesis from the second postnatal day onwards (Miething 1993), a tendency which, according to the present results, continues in early puberty up to the final termination of Sertoli cell proliferation on day 17. Advanced pubertal and adult Sertoli cells thus represent a numerically constant, presumably irreversibly postmitotic cell population, which may decrease again in number due to age-related cell loss (Johnson et al. 1984).

Differing data are available concerning the hormonal conditions of Sertoli cell proliferation. Steinberger and Steinberger (1971) and Steinberger (1973) were not able, on the basis of either in vivo or in vitro experiments, to determine hormone-dependent alterations of DNA synthesis or of proliferative activity of these cells. However, according to more recent investigations, Sertoli cell proliferation is reputed to be stimulated by FSH (Griswold et al. 1977; Orth 1984; Almiron and Chemes 1988) and by FSH and activin (Boitani et al. 1995), respectively. Nevertheless, this modulating effect varies in an age-related manner and may induce prolongation of the proliferative period for only a few days beyond its physiological cessation (Griswold et al. 1977) or, according to the study by Boitani et al. (1995), may cause no prolongation at all.

A thorough assessment of the present labeling indices of Sertoli cells reveals a general decline in their mitotic activity between days 12 and 16, being largely independent of the developmental level reached by the associated germ cells. Only in tubule segments in which the leading germ cells reach the leptotene phase of meiosis (developmental step IX-XII.1, see Table 6) are Sertoli cells mitotically quiescent as early as days 15 and 16. This might, nevertheless at low labeling indices (ranging from 0.4 to 2.8% on these days), point to additional, locally operative inhibition in these areas of the seminiferous epithelium.

The significance of the characteristic waning of Sertoli cell proliferation during early pubertal establishment of spermatogenesis is at present still a matter of speculation. It is conceivable that possible mitoses of (late-)pubertal or adult Sertoli cells would at least partially be incompatible with their manifold, continuous functions throughout the cycle of the seminiferous epithelium. This concerns the complex external shape encompassing various germ cell types (Wong and Russell 1983; Hamasaki and Murakami 1986) as well as the cytoskeletal organization (Vogl et al. 1993) and membrane contact structures of Sertoli cells. In this context, the continuous maintenance of the Sertoli cell barrier is probably especially important. On the other hand, an additional aspect arising from the present results needs to be considered: Sertoli cells might be debarred from their own mitotic proliferation because of their specific position within the seminiferous epithelium involving immediate exposure to the "meiotic milieu" of the adluminal compartment of the epithelium (see Sect. 4.6.2).

4.8 Conclusions

The results and relationships discussed so far reveal that pubertal development of the seminiferous epithelium in principle follows the pattern of adult spermatogenesis. In the course of this period, however, the germ cells do not yet develop under the environmental conditions present in adult spermatogenesis in the form of the mature somatic cells and a balanced hormonal and paracrine stimulation. Both structural and biochemical factors accompanying germ cell differentiation only mature concomitantly with the developmental progression of pubertal germ cells. Due to the complex interdependence of germ cell and somatic cell development and function, even minor temporal irregularities of somatic cell maturation might, therefore, interfere with orderly germ cell differentiation.

The results of the present study reveal two such temporary developmental perturbations of the seminiferous epithelium. First, the partial desynchronization of pubertal germ cell associations, resulting from the temporary developmental arrest of spermatids prior to initial elongation, is the outcome of a delayed establishment of the Leydig cell enzyme 11β-HSD. Second, the localized appearance of specifically arrested germ cell mitoses between days 14 and 20 seems to be connected with the fact that the Sertoli cell barrier does not reach complete functional maturation until some time after the neighboring germ cells initially enter meiosis.

As well as specifying the role of Leydig cell 11β-HSD and adding a new aspect to the functions of the Sertoli cell barrier, these results also elucidate regulative mechanisms and cellular interrelations of adult spermatogenesis. They may thus contribute to an improved understanding of possible alterations or disturbances of adult mammalian spermatogenesis which occur in connection with impairment of the above somatic parameters.

A general effect on orderly testicular development, is due to the timely testicular descent, i.e., one that is completed prior to the time when germ cells initially enter meiosis. The resulting lowering of temperature by 3–5°C influences the maturation and functional state of Leydig cells and the seminiferous epithelium (Hagenäs and Ritzen 1976; Jegou et al. 1984; O'Shaughnessy and Sheffield 1991; Eskola et al. 1995) and, in addition, possibly also affects epididymal function (Bedford 1978).

In the rat, testes have descended into the scrotum by approximately day 15 (Wensing 1986). As there is little or no constriction of the inguinal ring after testis passage, the testes can change their position freely between the scrotum and the peritoneal cavity for at least several weeks (Gier and Marion 1970; Wensing 1986). A comparable situation is present in the golden hamster. The position of the testis appears to vary between the inguen and the scrotum between day 20 and at least day 34, and the left and right testis of one animal may well occupy different positions (A. Miething, unpublished observation). An immediate correlation between testicular position and temporally constant, defined developmental events is, therefore, rather unlikely.

Studies on cell kinetics have determined a shorter duration of initial spermatogenesis compared with adult spermatogenesis in the mouse, rat, and Djungarian and Chinese hamsters (Kluin et al. 1982; van Haaster and de Rooij 1993). A similarly accelerated progression of pubertal spermatogenesis can be inferred from the present morphological data for the golden hamster (see Sect. 3.2.4.1). This is in agreement with the results obtained by Huckins (1965, 1973), who, on the basis of morphological and cell kinetic observations, also determined an acceleration of initial rat spermatogenesis and attributed this phenomenon to shortened generation times of differentiating spermatogonia and preleptotene spermatocytes.

Apart from the application of certain toxicants (Rosiepen et al. 1995), long-term alteration of testicular temperature (Meistrich et al. 1973) is the only factor able to influence the otherwise (species specifically) largely constant duration of spermatogenesis (Oakberg 1956b; Clermont and Harvey 1965; Go et al. 1971; Rosiepen et al. 1994). Both the experimentally demonstrated temperature dependency and the approximate temporal coincidence of testicular descent and deceleration of spermatogenesis to adult-like values (van Haaster and de Rooij 1993) point to a possible (direct or indirect) relationship between testicular descent and establishment of adult duration of spermatogensis.

5 Summary

The present study deals with initial spermatogenesis in the pubertal golden hamster (*Mesocricetus auratus*), in the course of which both the complex process of germ cell proliferation and differentiation and the cyclic pattern of organization of spermatogenesis are established.

Germ cells pass through the complete sequence of spermatogenic differentiation from the A-spermatogonium up to the mature spermatid between days 12 and 38. During this period, germ cell development is already based on the pattern of organization and synchronization characteristic of adult spermatogenesis:

1. Judged by the germ cell types present, each of the successively appearing germ cell associations conforms to one of the adult stages of the seminiferous epithelium (stage conformity of the developmental steps of the pubertal seminiferous epithelium).
2. The overall intratesticular variance of the developmental levels reached in different tubular segments each covers almost exactly one cycle length of spermatogenesis.
3. There is a regular, continuous succession of developmental steps along the longitudinal course of the tubules which, in every single tubule, comprises the interval of one spermatogenic cycle ("wave" of the pubertal seminiferous epithelium).
4. Taken together, the latter two results indicate that the tubules of one given testis are equal to one another with respect to their developmental level, i.e., they develop synchronously.

Germ cell propagation follows the clonal mode of proliferation. In quantitative terms, stage-related S-phase indices largely parallel the values of adult spermatogenesis. Both the frequently found germ cell apoptoses and degenerating intrameiotic spermatocytes correspond to degenerative phenomena in the adult seminiferous epithelium with respect to degenerative mode and stage-related occurrence.

Germ cell development is closely related to the maturation of both Sertoli cells and Leydig cells. At two developmental phases, localized disturbances of germ cell differentiation occur which are specific to the pubertal period and which correlate with a maturational delay of Sertoli cells and Leydig cells, respectively:

1. Completion of Sertoli cell barrier formation may temporally follow the initial onset of meiosis in the neighboring germ cells. As a result, mitotically dividing germ cells fall under the influence of a (hitherto hypothetic) "meiotic milieu" and characteristically degenerate (arrested germ cell mitoses). This relationship emphasizes the functional significance of the separation of the two seminiferous compartments and, in addition, heightens the understanding of the necessity of normal termination of Sertoli cell proliferation before germ cells initially enter meiosis.

2. The Leydig cell enzyme 11β-HSD, which is known to promote testosterone output, is established several days after the leading spermatids initially reach step 6 of spermiogenesis (which is immediately prior to the particularly testosterone-dependent start of elongation). The resulting selective developmental arrest of round spermatids of steps 6–6/7 between days 26 and 30 leads to a localized, partial desynchronization of the seminiferous epithelium, which can be traced for several days after the above-mentioned maturational delay of Leydig cells has been overcome and which is accompanied by an extensive degenerative disposition of the spermatids involved.

The present investigation documents an organizational and synchronizational level of initial spermatogenesis that is similar to the situation in the adult animal. The analysis of temporarily occurring perturbations in the process of establishing the pubertal seminiferous epithelium emphasizes and specifies the close structural and functional relationships between the maturational state of somatic cells and germ cell differentiation.

References

Allan DJ, Harmon BV, Kerr JFR (1987) Cell death in spermatogenesis. In: Potten CS (ed) Perspectives of mammalian cell death. Oxford University Press, p 229–258

Allan DJ, Harmon BV, Roberts SA (1992) Spermatogonial apoptosis has three morphologically recognizable phases and shows no circadian rhythm during normal spermatogenesis in the rat. Cell Prolif 25: 241–250

Allen E, Altland PD (1952) Studies on degenerating sex cells in immature mammals. II. Modes of degeneration in the normal differentiation of the definitive germ cells in the male albino rat from age twelve days to maturity. J Morphol 91: 515–539

Almiron I, Chemes H (1988) Spermatogenic onset. II. FSH modulates mitotic activity of germ and Sertoli cells in immature rats. Int J Androl 11: 235–246

Amlani S, Vogl AW (1988) Changes in the distribution of microtubules and intermediate filaments in mammalian Sertoli cells during spermatogenesis. Anat Rec 220: 143–160

Awoniyi CA, Santulli R, Sprando RL, Ewing LL, Zirkin BR (1989a) Restoration of advanced spermatogenic cells in the experimentally regressed rat testis: quantitative relationship to testosterone concentration within the testis. Endocrinology 124: 1217–1223

Awoniyi CA, Santulli R, Chandrashekar V, Schanbacher BD, Zirkin BR (1989b) Quantitative restoration of advanced spermatogenic cells in adult male rats made azoospermic by active immunization against luteinizing hormone or gonadotropin-releasing hormone. Endocrinology 125: 1303–1309

Baker TG (1963) A quantitative and cytological study of germ cells in human ovaries. Proc R Soc Lond (Biol) 158: 417–433

Baker TG, Franchi LL (1967) The fine structure of oogonia and oocytes in human ovaries. J Cell Sci 2: 213–224

Barr AB, Moore DJ, Paulsen CA (1971) Germinal cell loss during human spermatogenesis. J Reprod Fert 25: 75–80

Bartlett JMS, Kerr JB, Sharpe RM (1986) The effect of selective destruction and regeneration of rat Leydig cells on the intratesticular distribution of testosterone and morphology of the seminiferous epithelium. J Androl 7: 240–253

Bartlett JMS, Kerr JB, Sharpe RM (1988) The selective removal of pachytene spermatocytes using methoxy acetic acid as an approach to the study in vivo of paracrine interactions in the testis. J Androl 9: 31–40

Bartlett JMS, Weinbauer GF, Nieschlag E (1989) Differential effects of FSH and testosterone on the maintenance of spermatogenesis in the adult hypophysectomized rat. J Endocrinol 121: 49–58

Beaumont HM, Mandl AM (1962) A quantitative and cytological study of oogonia and oocytes in the foetal and neonatal rat. Proc R Soc Lond (Biol) 155: 557–579

Bedford JM (1978) Anatomical evidence for the epididymis as the prime mover in the evolution of the scrotum. Am J Anat 152: 483–508

Benda C (1887) Untersuchungen über den Bau des funktionirenden Samenkanälchens einiger Säugethiere und Folgerungen für die Spermatogenese dieser Wirbelthierklasse. Arch Mikr Anat 30: 49–110

Bergmann M (1987) Photoperiod and testicular function in Phodopus sungorus. In: Beck F, Hild W, Kriz W, Ortmann R, Pauly JE, Schiebler TH (eds) Advances in anatomy, embryology and cell biology, vol 105. Springer, Berlin Heidelberg New York

Bergmann M, Dierichs R (1983) Postnatal formation of the blood-testis barrier in the rat with special reference to the initiation of meiosis. Anat Embryol 168: 269–275
Billig H, Furuta I, Rivier C, Tapanainen J, Parvinen M, Hsueh AJW (1995) Apoptosis in testis germ cells: developmental changes in gonadotropin dependence and localization to selective tubule stages. Endocrinology 136: 5–12
Boitani C, Stefanini M, Fragale A, Morena AR (1995) Activin stimulates Sertoli cell proliferation in a defined period of rat testis development. Endocrinology 136: 5438–5444
Boyer CC (1953) Chronology of development for the golden hamster. J Morphol 92: 1–38
Braun RE, Behringer RR, Peschon JJ, Brinster RL, Palmiter RD (1989) Genetically haploid spermatids are phenotypically diploid. Nature 337: 373–376
Brinkworth MH, Weinbauer GF, Schlatt S, Nieschlag E (1995) Identification of male germ cells undergoing apoptosis in adult rats. J Reprod Fert 105: 25–33
Brown HH (1885) On spermatogenesis in the rat. Quart J Micr Sci 25: 343–369
Bustos-Obregon E, Castellon E, Balbontin J (1988) Spermatogonial chalones. In: Holstein AF, Leidenberger F, H"lzer KH, Bettendorf G (eds) Carl Schirren symposium: advances in andrology. Diesbach, Berlin, p 53–58
Byers S, Pelletier RM, SuarezQuian C (1993) Sertoli cell junctions and the seminiferous epithelium barrier. In: Russell LD, Griswold MD (eds) The Sertoli cell. Cache River, Clearwater, p 431–446
Cameron DF, Muffly KE (1991) Hormonal regulation of spermatid binding. J Cell Sci 100: 623–633
Cameron DF, Muffly KE, Nazian SJ (1993) Reduced testosterone during puberty results in a midspermiogenic lesion. Proc Soc Exp Biol Med 202: 457–464
Castellon E, Janecki A, Steinberger A (1989) Influence of germ cells on Sertoli cell secretory activity in direct and indirect coculture with Sertoli cells from rats of different ages. Mol Cell Endocrinol 64: 169–178
Challoner S (1974) Studies of oogenesis and follicular development in the golden hamster. 1. A quantitative study of meiotic prophase in vivo. J Anat 117: 373–383
Chemes H (1986) The phagocytic function of Sertoli cells: a morphological, biochemical, and endocrinological study of lysosomes and acid phosphatase localization in the rat testis. Endocrinology 119: 1673–1681
Chowdhury AW, Steinberger E (1964) A quantitative study of the effect of heat on germinal epithelium of rat testes. Am J Anat 115: 509–524
Clemens LG, Witcher JA (1985) Sexual differentiation and development. In: Siegel HI (ed) The hamster. Reproduction and behavior. Plenum, New York, p 155–171
Clermont Y (1954) Cycle de l'epithelium seminal et mode de renouvellement des spermatogonies chez le hamster. Rev Canad Biol 13: 208–245
Clermont Y (1962) Quantitative analysis of spermatogenesis of the rat: a revised model for the renewal of spermatogonia. Am J Anat 111: 111–129
Clermont Y (1972) Kinetics of spermatogenesis in mammals: seminiferous epithelium cycle and spermatogonial renewal. Physiol Rev 52: 198–236
Clermont Y, Harvey SC (1965) Duration of the cycle of the seminiferous epithelium of normal, hypophysectomized and hypophysectomized-hormone treated albino rats. Endocrinology 76: 80–89
Clermont Y, Huckins C (1961) Microscopic anatomy of the sex cords and seminiferous tubules in growing and adult male albino rats. Am J Anat 108: 79–97
Clermont Y, Mauger A (1974) Existence of a spermatogonial chalone in the rat testis. Cell Tissue Kinet 7: 165–172
Clermont Y, Morgentaler H (1955) Quantitative study of spermatogenesis in the hypophysectomized rat. Endocrinology 57: 369–382
Clermont Y, Perey B (1957a) Quantitative study of the cell population of the seminiferous tubules in immature rats. Am J Anat 100: 241–268
Clermont Y, Perey B (1957b) The stages of the cycle of the seminiferous epithelium of the rat: practical definitions in PA-Schiff-hematoxylin and hematoxylin-eosin stained sections. Rev Canad Biol 16: 451–462
Clermont Y, Trott M (1969) Duration of the cycle of the seminiferous epithelium in the mouse and hamster determined by means of 3H-thymidine and radioautography. Fertil Steril 20: 805–817
Connell CJ (1974) The Sertoli cell of the sexually mature dog. Anat Rec 178: 333

Connell CJ (1980) Blood-testis barrier formation and the initiation of meiosis in the dog. In: Steinberger A, Steinberger E (eds) Testicular development, structure, and function. Raven, New York, p 71–78

Cordell JL, Falini B, Erber WN, Ghosh AK, Abdulaziz Z, MacDonald S, Pulford KAF, Stein H, Mason DY (1984) Immunoenzymatic labeling of monoclonal antibodies using immune complexes of alkaline phosphatase and monoclonal anti-alkaline phosphatase (APAAP complexes). J Histochem Cytochem 32: 219–229

Cunningham GR, Huckins C (1979a) Failure to identify a spermatogonial chalone in adult irradiated testes. Cell Tissue Kinet 9: 81–89

Cunningham GR, Huckins C (1979b) Persistence of complete spermatogenesis in the presence of low intratesticular concentrations of testosterone. Endocrinology 105: 177–186

DeFelipe J, Fairen A (1993) A simple and reliable method for correlative light and electron microscopic studies. J Histochem Cytochem 41: 769–772

de Jong FH (1988) Inhibin. Physiol Rev 68: 555–607

de Rooij DG (1968) Stem cell renewal and duration of spermatogonial cycle in the goldhamster. Z Zellforsch 89: 133–136

de Rooij DG, Lok D (1987) Regulation of the density of spermatogonia in the seminiferous epithelium of the Chinese hamster. II. Differentiating spermatogonia. Anat Rec 217: 131–136

Dietert SE (1966) Fine structure of the formation and fate of the residual bodies of mouse spermatozoa with evidence for the participation of lysosomes. J Morphol 120: 317–346

Dolbeare F (1995) Bromodeoxyuridine: a diagnostic tool in biology and medicine. 1. Historical perspectives, histochemical methods and cell kinetics. Histochem J 27: 339–369

Drenckhahn D, Franz H (1986) Identification of actin-, α-actinin, and vinculin-containing plaques at the lateral membrane of epithelial cells. J Cell Biol 102: 1843–1852

Dustin P (1984) Microtubules, 2nd edn. Springer, Berlin Heidelberg New York

Dym M, Clermont Y (1970) Role of spermatogonia in the repair of the seminiferous epithelium following X-irradiation of the rat testis. Am J Anat 128: 265–282

Dym M, Fawcett DW (1970) The blood-testis barrier in the rat and the physiological compartmentation of the seminiferous epithelium. Biol Reprod 3: 308–326

Eskola V, Paukku T, Warren DW, Huhtaniemi I (1995) A novel role for testicular descent; temperature-dependent induction of pertussis toxin-sensitive Gi protein function in postnatal rat Leydig cells. Endocrinology 136: 4659–4664

Fawcett DW (1961) Intercellular bridges. Exp Cell Res [Suppl] 8: 174–187

Fawcett DW (1975) Ultrastructure and function of the Sertoli cell. In: Hamilton DW, Greep RO (eds) Handbook of physiology, sect 7, vol 5. Male reproductive system. American Physiological Society, Washington, p 21–55

Fawcett DW, Ito S, Slautterback D (1959) The occurrence of intercellular bridges in groups of cells exhibiting synchronous differentiation. J Biophys Biochem Cytol 5: 453–460

Fawcett DW, Leak LC, Heidger PM (1970) Electron microscopic observations on the structural components of the blood-testis barrier. J Reprod Fert [Suppl] 10: 105–122

Franchi LL, Mandl AM (1963) The ultrastructure of oogonia and oocytes in the foetal and neonatal rat. Proc R Soc Lond (Biol) 157: 99–114

Frungieri MB, Gonzalez-Calvar SI, Chandrashekar V, Rao JN, Bartke A, Calandra RS (1996) Testicular gamma-aminobutyric acid and circulating androgens in Syrian and Djungarian hamsters during sexual development. Int J Androl 19: 164–170

Furuchi T, Masuko K, Nishimune Y, Obinata M, Matsui Y (1996) Inhibition of testicular germ cell apoptosis and differentiation in mice misexpressing Bcl-2 in spermatogonia. Development 122: 1703–1709

Galdieri M, Monaco L, Stefanini M (1984) Secretion of androgen-binding protein by Sertoli cells is influenced by contact with germ cells. J Androl 5: 409–415

Gao H-B, Ge R-S, Lakshmi V, Marandici A, Hardy MP (1997) Hormonal regulation of oxidative and reductive activities of 11β-hydroxysteroid dehydrogenase in rat Leydig cells. Endocrinology 138: 156–161

Ghosh S, Sinha Hikim AP, Russell LD (1991) Further observations of stage-specific effects seen after short-term hypophysectomy in the rat. Tissue Cell 23: 613–630

Gier HT, Marion GB (1970) Development of the mammalian testis. In: Johnson AD, Gomes WR, Vandemark NL (eds) The testis. I. Development, anatomy, and physiology. Academic, New York, p 1–45
Gilula NB, Fawcett DW, Aoki A (1976) The Sertoli cell occluding junctions and gap junctions in mature and developing mammalian testis. Dev Biol 50: 142–168
Go VLW, Vernon RG, Fritz IB (1971) Studies on spermatogenesis in rats. III. Effects of hormonal treatment on differentiation kinetics of the spermatogenic cycle in regressed hypophysectomized rats. Can J Biochem 49: 768–775
Gondos B (1973) Intercellular bridges and mammalian germ cell differentiation. Differentiation 1: 177–182
Gondos B (1984) Germ cell differentiation and intercellular bridges. In: Van Blerkom J, Motta PM (eds) Ultrastructure of reproduction. Nijhoff, Dordrecht, p 31–45
Gondos B, Berndtson WE (1993) Postnatal and pubertal development. In: Russell LD, Griswold MD (eds) The Sertoli cell. Cache River, Clearwater, p 115–154
Gondos B, Paup DC, Ross J, Gorski RA (1974) Ultrastructural differentiation of Leydig cells in the fetal and postnatal hamster testis. Anat Rec 178: 551–566
Gratzner HG (1982) Monoclonal antibody to 5-bromo- and 5-iododeoxyuridine: a new reagent for detection of DNA replication. Science 218: 474–475
Griswold MD (1988) Protein secretions of Sertoli cells. Int Rev Cytol 110: 133–156
Griswold MD (1993) Protein secretion by Sertoli cells: general considerations. In: Russell LD, Griswold MD (eds) The Sertoli cell. Cache River, Clearwater, p 195–200
Griswold MD, Solari A, Tung PS, Fritz IB (1977) Stimulation by follicle-stimulating hormone of DNA synthesis and of mitosis in cultured Sertoli cells prepared from testes of immature rats. Mol Cell Endocrinol 7: 151–165
Hagenäs L, Ritzen EM (1976) Impaired Sertoli cell function in experimental cryptorchidism in the rat. Mol Cell Endocrinol 4: 25–34
Hagenäs L, Plöen L, Ekwall H (1978) Blood-testis barrier: evidence for intact inter-Sertoli cell junctions after hypophysectomy in the adult rat. J Endocrinol 76: 87–91
Haider SG (1988) Leydigzellen. Funktionelle Morphologie und Enzymhistochemie bei Ratte und Mensch. Thieme, Stuttgart
Haider SG, Berthold G, Servos G (1995) Androgenrezeptor in den männlichen Genitalorganen der Marmosetaffen (Callithrix jacchus); eine immunhistochemische Studie. Ann Anat [Suppl] 177: 109–110
Hamasaki M, Murakami M (1986) Three-dimensional profiles of Sertoli cell processes and associated appendices as viewed by SEM. J Electron Microsc 35: 132–143
Henriksen K, Kulmala J, Toppari J, Mehrotra K, Parvinen M (1996) Stage-specific apoptosis in the rat seminiferous epithelium: quantification of irradiation effects. J Androl 17: 394–402
Hilscher W (1983) Problems of the Keimbahn. Biblthca Anat 24: 1–21
Hilscher W, Makoski HB (1968) Histologische und autoradiographische Untersuchungen zur "Präspermatogenese" und "Spermatogenese" der Ratte. Z Zellforsch 86: 327–350
Hilscher B, Hilscher W, Bülthoff-Ohnolz B, Krämer U, Birke A, Pelzer H, Gauss G (1974) Kinetics of gametogenesis. I. Comparative histological and autoradiographic studies of oocytes and transitional prospermatogonia during oogenesis and prespermatogenesis. Cell Tissue Res 154: 443–470
Holstein AF, Wulfhekel U (1971) Die Semidünnschnitt-Technik als Grundlage für eine cytologische Beurteilung der Spermatogenese des Menschen. Andrologie 3: 65–69
Horan AH, Bedford JM (1972) Development of the fertilizing ability of spermatozoa in the epididymis of the Syrian hamster. J Reprod Fert 30: 417–423
Huckins C (1965) Duration of spermatogenesis in pre- and postpuberal Wistar rats. Anat Rec 151: 364
Huckins C (1972) The morphology and kinetics of spermatogonial degeneration in normal adult rats. Biol Reprod 7: 131–132
Huckins C (1973) Cell cycle properties of differentiating spermatogonia and primary spermatocytes in immature Sprague-Dawley rats. Anat Rec 175: 347–348
Huckins C (1978) The morphology and kinetics of spermatogonial degeneration in adult rats: an analysis using a simplified classification of the germinal epithelium. Anat Rec 190: 905–926
Ito S, Winchester RJ (1963) The fine structure of the gastric mucosa in the bat. J Cell Biol 16: 541–577
Jegou B (1993) The Sertoli-germ cell communication network in mammals. Int Rev Cytol 147: 25–96

Jegou B, Peake RA, Irby DC, de Kretser DM (1984) Effects of the induction of experimental cryptorchidism and subsequent orchidopexy on testicular function in immature rats. Biol Reprod 30: 179–187

Johnson MH (1970) An immunological barrier in the guinea-pig testis. J Pathol 101: 129–139

Johnson L, Zane RS, Petty CS, Neaves WB (1984) Quantification of the human Sertoli cell population: its distribution, relation to germ cell numbers, and age-related decline. Biol Reprod 31: 785–795

Karnovsky MJ (1965) A formaldehyde-glutaraldehyde fixative of high osmolality for use in electron microscopy. J Cell Biol 27: 137A–138A

Kerr JB (1992) Spontaneous degeneration of germ cells in normal rat testis: assessment of cell types and frequency during the spermatogenic cycle. J Reprod Fert 95: 825–830

Kerr JB, de Kretser DM (1974) The role of the Sertoli cell in phagocytosis of the residual bodies of spermatids. J Reprod Fert 36: 439–440

Kerr JFR, Wyllie AH, Currie AR (1972) Apoptosis: a basic biological phenomenon with wide-ranging implications in tissue kinetics. Br J Cancer 26: 239–257

Kluin PM, de Rooij DG (1981) A comparison between the morphology and cell kinetics of gonocytes and adult type undifferentiated spermatogonia in the mouse. Int J Androl 4: 475–493

Kluin PM, Kramer MF, de Rooij DG (1982) Spermatogenesis in the immature mouse proceeds faster than in the adult. Int J Androl 5: 282–294

Kluin PM, Kramer MF, de Rooij DG (1984) Proliferation of spermatogonia and Sertoli cells in maturing mice. Anat Embryol 169: 73–78

Knorr DW, Vanha-Perttula T, Lipsett MB (1970) Structure and function of rat testis through pubescence. Endocrinology 86: 1298–1304

Larramendi PCH (1985) Method of retrieval of one to two micron sections from glass for ultrathin sectioning. J Electron Microsc Tech 2: 645–646

Leblond CP, Clermont Y (1952) Definition of the stages of the cycle of the seminiferous epithelium in the rat. Ann NY Acad Sci 55: 548–573

Le Magueresse B, Jegou B (1988) Paracrine control of immature Sertoli cells by adult germ cells, in the rat (an in vitro study). Cell-cell interactions within the testis. Mol Cell Endocrinol 58: 65–72

Lojda Z (1965) Remarks on histochemical demonstration of dehydrogenases. II. Intracellular localization. Folia Morphol (Prague) 13: 84–96

Lojda Z, Gossrau R, Schiebler TH (1979) Enzyme histochemistry: a laboratory manual. Springer, Berlin Heidelberg New York

MacCalman CD, O'Brien DA, Byers S, Blaschuk OW (1993) N-cadherin expression in the seminiferous epithelium of the mouse testis. Endocrine J 1: 519–525

Maunsbach AB (1978) Electron microscopic analysis of objects in light microscopic sections. In: Sturgess JM, Kalnins VI, Ottensmeyer FP, Simon GT (eds) Proceedings of the 9th International Congress on Electron Microscopy, Toronto, vol II. Imperial, Ontario, p 80–81

Mayor HD, Hampton JC, Rosario B (1961) A simple method for removing the resin from epoxy-embedded tissue. J Biophys Biochem Cytol 9: 909–910

Means AR, Fakunding JL, Huckins C, Tindall DJ, Vitale R (1976) Follicle-stimulating hormone, the Sertoli cell, and spermatogenesis. Recent Progr Horm Res 32: 477–522

Meistrich ML, Eng VWS, Loir M (1973) Temperature effects on the kinetics of spermatogenesis in the mouse. Cell Tissue Kinet 6: 379–393

Miething A (1989) Morphological studies on prespermatogonia and pre-Sertoli cells in the testes of 6- to 11-day-old golden hamsters. Anat Embryol 179: 503–510

Miething A (1990) Intercellular bridges between germ cells in the immature golden hamster testis: evidence for clonal and non-clonal mode of proliferation. Cell Tissue Res 262: 559–567

Miething A (1992a) Germ-cell death during prespermatogenesis in the testis of the golden hamster. Cell Tissue Res 267: 583–590

Miething A (1992b) Ultrathin sectioning of different areas of the same semithin section. Microsc Res Tech 21: 73–74

Miething A (1993) Proliferative activity of the developing seminiferous epithelium during prespermatogenesis in the golden hamster testis measured by bromodeoxyuridine labeling. Anat Embryol 187: 249–258

Miething A (1995) The bridge-partitioning complex of germ-cell intercellular bridges in the testis of the golden hamster. Cell Tissue Res 281: 359–365

Miething A (1996) Flat plastic embedding and precise longitudinal sectioning of isolated testicular seminiferous tubules. Andrologia 28: 343–347

Miething A (1997) Degeneration of spermatocytes during meiotic divisions in the golden hamster testis. J Submicrosc Cytol Pathol 29: 29–35

Miller LL, Whitsett JM, Vandenbergh JG, Colby DR (1977) Physical and behavioral aspects of sexual maturation in male golden hamsters. J Comp Physiol Psychol 91: 245–259

Monder C, Shackleton CHL (1984) 11β-Hydroxysteroid dehydrogenase: fact or fancy? Steroids 44: 383–417

Monder C, Sakai RR, Miroff Y, Blanchard DC, Blanchard RJ (1994a) Reciprocal changes in plasma corticosterone and testosterone in stressed male rats maintained in a visible burrow system: evidence for a mediating role of testicular 11β-hydroxysteroid dehydrogenase. Endocrinology 134: 1193–1198

Monder C, Miroff Y, Marandici A, Hardy MP (1994b) 11β-Hydroxysteroid dehydrogenase alleviates glucocorticoid-mediated inhibition of steroidogenesis in rat Leydig cells. Endocrinology 134: 1199–1204

Morales C, Clermont Y (1993) Structural changes of the Sertoli cell during the cycle of the seminiferous epithelium. In: Russell LD, Griswold MD (eds) The Sertoli cell. Cache River, Clearwater, p 305–329

Mori C, Nakamura N, Dix DJ, Fujioka M, Nakagawa S, Shiota K, Eddy EM (1997) Morphological analysis of germ cell apoptosis during postnatal testis development in normal and Hsp70-2 knockout mice. Dev Dyn 208: 125–136

Muffly KE, Nazian SJ, Cameron DF (1994) Effects of follicle-stimulating hormone on the junction-related Sertoli cell cytoskeleton and daily sperm production in testosterone-treated hypophysectomized rats. Biol Reprod 51: 158–166

Nagano T, Suzuki F (1976) The postnatal development of the junctional complexes of the mouse Sertoli cells as revealed by freeze-fracture. Anat Rec 185: 403–418

Neumann A, Haider SG, Hilscher B (1993) Temporal coincidence of the appearance of elongated spermatids and of histochemical reaction of 11β-hydroxysteroid dehydrogenase in rat Leydig cells. Andrologia 25: 263–269

Newton SC, Blaschuk OW, Millette CF (1993) N-Cadherin mediates Sertoli cell-spermatogenic cell adhesion. Dev Dyn 197: 1–13

Oakberg EF (1956a) A description of spermiogenesis in the mouse and its use in analysis of the cycle of the seminiferous epithelium and germ cell renewal. Am J Anat 99: 391–413

Oakberg EF (1956b) Duration of spermatogenesis in the mouse and timing of stages of the cycle of the seminiferous epithelium. Am J Anat 99: 507–516

O'Donnell L, McLachlan RI, Wreford NG, Robertson DM (1994) Testosterone promotes the conversion of round spermatids between stages VII and VIII of the rat spermatogenic cycle. Endocrinology 135: 2608–2614

Onoda M, Djakiew D (1991) Pachytene spermatocyte protein(s) stimulate Sertoli cells grown in bicameral chambers: dose-dependent secretion of ceruloplasmin, sulfated glycoprotein-1, sulfated glycoprotein-2, and transferrin. In Vitro Cell Dev Biol 27A: 215–222

Orth JM (1982) Proliferation in Sertoli cells in fetal and postnatal rats: a quantitative autoradiographic study. Anat Rec 203: 485–492

Orth JM (1984) The role of follicle-stimulating hormone in controlling Sertoli cell proliferation in testes of fetal rats. Endocrinology 115: 1248–1255

Orth J, Christensen AK (1977) Localization of 125I-labeled FSH in the testes of hypophysectomized rats by autoradiography at the light and electron microscope levels. Endocrinology 101: 262–278

Ortlip SA, Li SA, Li JJ (1981) Characterization of specific glucocorticoid receptor in the Syrian hamster testis. Endocrinology 109: 1331–1338

O'Shaughnessy PJ, Sheffield JW (1991) Effect of temperature and the role of testicular descent on post-natal testicular androgen production in the mouse. J Reprod Fert 91: 357–364

Parvinen M (1993) Cyclic function of Sertoli cells. In: Russell LD, Griswold MD (eds) The Sertoli cell. Cache River, Clearwater, p 331–347

Pelletier RM, Byers SW (1992) The blood-testis barrier and Sertoli cell junctions: structural considerations. Microsc Res Tech 20: 3–33

Pelletier RM, Friend DS (1983) The Sertoli cell junctional complex: structure and permeability to filipin in the neonatal and adult guinea pig. Am J Anat 168: 213–228

Perey B, Clermont Y, Leblond CP (1961) The wave of the seminiferous epithelium in the rat. Am J Anat 108: 47–77

Peter K (1899) Die Bedeutung der Nährzelle im Hoden. Arch Mikr Anat 53: 180–211

Phillips DM, Lakshmi V, Monder C (1989) Corticosteroid 11β-dehydrogenase in rat testis. Endocrinology 125: 209–216

Pineau C, Le Magueresse B, Courtens JL, Jegou B (1991) Study in vitro of the phagocytic function of Sertoli cells in the rat. Cell Tissue Res 264: 589–598

Price D, Ortiz E (1965) The role of fetal androgen in sex differentiation in mammals. In: De Haan RL, Ursprung H (eds) Organogenesis. Holt, Rinehart and Winston, New York, p 629–652

Pützer B (1986) Histologische Untersuchungen an Semidünnschnitten über die Gonadenentwicklung beim Goldhamster und Vergleich der Goniogenese mit anderen Säugertypen und Mensch. Dissertation, Faculty of Biology, University of Bonn

Regaud C (1900) Degenerescence des cellules seminales chez les mammiferes en l'absence de tout etat pathologique. C R Soc Biol 52: 268–270

Regaud C (1901) Etudes sur la structure des tubes seminiferes et sur la spermatogenese chez les mammiferes. Arch Anat Micr 4: 101–105, 231–380

Ren HP, Russell LD (1991) Clonal development of interconnected germ cells in the rat and its relationship to the segmental and subsegmental organization of spermatogenesis. Am J Anat 192: 121–128

Reynolds ES (1963) The use of lead citrate at high pH as an electron opaque stain in electron microscopy. J Cell Biol 17: 208–212

Robb GW, Amann RP, Killian GJ (1978) Daily sperm production and epididymal sperm reserves of pubertal and adult rats. J Reprod Fert 54: 103–107

Rommerts FFG (1988) How much androgen is required for maintenance of spermatogenesis? J Endocrinol 116: 7–9

Roosen-Runge EC (1952) Kinetics of spermatogenesis in mammals. Ann NY Acad Sci 55: 574–584

Roosen-Runge EC (1955) Untersuchungen über die Degeneration samenbildender Zellen in der normalen Spermatogenese der Ratte. Z Zellforsch 41: 221–235

Roosen-Runge EC (1973) Germinal-cell loss in normal metazoan spermatogenesis. J Reprod Fert 35: 339–348

Roosen-Runge EC (1974) Die Spermatogenese im Lichte der Evolution. Anat Anz [Suppl] 136: 23–37

Roosen-Runge EC, Giesel LO (1950) Quantitative studies on spermatogenesis in the albino rat. Am J Anat 87: 1–30

Rosiepen G, Weinbauer GF, Schlatt S, Behre HM, Nieschlag E (1994) Duration of the cycle of the seminiferous epithelium, estimated by the 5-bromodeoxyuridine technique, in laboratory and feral rats. J Reprod Fert 100: 299–306

Rosiepen G, Chapin RE, Weinbauer GF (1995) The duration of the cycle of the seminiferous epithelium is altered by administration of 2,5-hexanedione in the adult Sprague-Dawley rat. J Androl 16: 127–135

Russell L (1977a) Movement of spermatocytes from the basal to the adluminal compartment of the rat testis. Am J Anat 148: 313–328

Russell LD (1977b) Observations on rat Sertoli ectoplasmic ("junctional") specializations in their association with germ cells of the rat testis. Tissue Cell 9: 475–498

Russell LD (1978) The blood-testis barrier and its formation relative to spermatocyte maturation in the adult rat: a lanthanum tracer study. Anat Rec 190: 99–112

Russell LD (1980) Sertoli-germ cell interrelations: a review. Gamete Res 3: 179–202

Russell LD (1993) Morphological and functional evidence for Sertoli-germ cell relationships. In: Russell LD, Griswold MD (eds) The Sertoli cell. Cache River, Clearwater, p 365–390

Russell LD, Clermont Y (1976) Anchoring device between Sertoli cells and late spermatids in rat seminiferous tubules. Anat Rec 185: 259–278

Russell LD, Clermont Y (1977) Degeneration of germ cells in normal, hypophysectomized and hormone treated hypophysectomized rats. Anat Rec 187: 347–366

Russell LD, Griswold MD (eds) (1993) The Sertoli cell. Cache River, Clearwater

Russell LD, Malone JP, Karpas SL (1981) Morphological pattern elicited by agents affecting spermatogenesis by disruption of its hormonal stimulation. Tissue Cell 13: 369–380

Russell LD, Alger LE, Nequin LG (1987) Hormonal control of pubertal spermatogenesis. Endocrinology 120: 1615–1632
Russell LD, Bartke A, Goh JC (1989) Postnatal development of the Sertoli cell barrier, tubular lumen, and cytoskeleton of Sertoli and myoid cells in the rat, and their relationship to tubular fluid secretion and flow. Am J Anat 184: 179–189
Russell LD, Ettlin RA, Sinha Hikim AP, Clegg ED (1990) Histological and histopathological evaluation of the testis. Cache River, Clearwater
Salanova M, Stefanini M, De Curtis I, Palombi F (1995) Integrin receptor α6β1 is localized at specific sites of cell-to-cell contact in rat seminiferous epithelium. Biol Reprod 52: 79–87
Sanchez B, Pizarro M, Garcia P, Flores JM (1993) Postnatal development of seminiferous tubules in the cat. J Reprod Fert [Suppl] 47: 343–348
Schabtach E, Parkening TA (1974) A method for sequential high resolution light and electron microscopy of selected areas of the same material. J Cell Biol 61: 261–264
Schubert W (1977) Lichtmikroskopische Untersuchungen über die Beeinflussung der Spermatogenese durch Lichtentzug beim Goldhamster (Mesocricetus auratus). Dissertation, Faculty of Medicine, University of Bonn
Schulze C (1984) Sertoli cells and Leydig cells in man. In: Beck F, Hild W, Ortmann R, Pauly JE, Schiebler TH (eds) Advances in anatomy, embryology and cell biology, vol 88. Springer, Berlin Heidelberg New York
Schulze W, Rehder U (1984) Organization and morphogenesis of the human seminiferous epithelium. Cell Tissue Res 237: 395–407
Schulze W, Riemer M, Rehder U, Höhne KH (1986) Computer-aided three-dimensional reconstructions of the arrangement of primary spermatocytes in human seminiferous tubules. Cell Tissue Res 244: 1–8
Sertoli E (1865) Dell'esistenza di particolari cellule ramificate nei canalicoli seminiferi del testicolo umano. Morgagni 7: 31–40
Setchell BP (1980) The functional significance of the blood-testis barrier. J Androl 1: 3–10
Sharpe RM (1986) Paracrine control of the testis. Clin Endocrinol Metab 15: 185–207
Sharpe RM (1987) Testosterone and spermatogenesis. J Endocrinol 113: 1–2
Sharpe RM, Fraser HM, Cooper I, Rommerts FF (1982) The secretion, measurement, and function of a testicular LHRH-like factor. Ann NY Acad Sci 383: 272–294
Sharpe RM, Donachie K, Cooper I (1988) Re-evaluation of the intratesticular level of testosterone required for quantitative maintenance of spermatogenesis in the rat. J Endocrinol 117: 19–26
Sinha Hikim AP, Wang C, Leung A, Swerdloff RS (1995) Involvement of apoptosis in the induction of germ cell degeneration in adult rats after gonadotropin-releasing hormone antagonist treatment. Endocrinology 136: 2770–2775
Skinner MK (1991) Cell-cell interactions in the testis. Endocrine Rev 12: 45–77
Skinner MK (1993) Secretion of growth factors and other regulatory factors. In: Russell LD, Griswold MD (eds) The Sertoli cell. Cache River, Clearwater, p 237–247
Spiteri-Grech J, Nieschlag E (1993) Paracrine factors relevant to the regulation of spermatogenesis – a review. J Reprod Fert 98: 1–14
Sprando RL (1990) Perfusion of the rat testis through the heart using heparin. In: Russell LD, Ettlin RA, Sinha Hikim AP, Clegg ED (eds) Histological and histopathological evaluation of the testis. Cache River, Clearwater, p 277–280
Stalker A, Hermo L, Antakly T (1989) Covalent affinity labeling, radioautography, and immunocytochemistry localize the glucocorticoid receptor in rat testicular Leydig cells. Am J Anat 186: 369–377
Steinberger A (1973) Effect of gonadotropins on the replication of Sertoli cells in immature and immature-hypophysectomized rats. Biol Reprod 9: 61–62
Steinberger A, Steinberger E (1971) Replication pattern of Sertoli cells in maturing rat testis in vivo and in organ culture. Biol Reprod 4: 84–87
Steinberger A, Steinberger E (1977) The Sertoli cells. In: Johnson AD, Gomes WR (eds) Advances in physiology, biochemistry, and function. IV. The testis. Academic, New York, p 371–399
Stempak JG, Ward RT (1964) An improved staining method for electron microscopy. J Cell Biol 22: 697–701

Sun EL, Gondos B (1981) Proliferative activity in the rabbit testis during postnatal development. In: Byskov AG, Peters H (eds) Development and function of reproductive organs. Excerpta Medica, Amsterdam, p 140–148

Sun EL, Gondos B (1986) Formation of the blood-testis barrier in the rabbit. Cell Tissue Res 243: 575–578

Sun YT, Wreford NG, Robertson DM, de Kretser DM (1990) Quantitative cytological studies of spermatogenesis in intact and hypophysectomized rats: identification of androgen-dependent stages. Endocrinology 127: 1215–1223

Swierstra EE, Foote RH (1963) Cytology and kinetics of spermatogenesis in the rabbit. J Reprod Fert 5: 309–322

Tapanainen JS, Tilly JL, Vihko KK, Hsueh AJW (1993) Hormonal control of apoptotic cell death in the testis: gonadotropins and androgens as testicular cell survival factors. Mol Endocrinol 7: 643–650

Tindall DJ, Miller DA, Means AR (1977) Characterization of androgen receptor in Sertoli cell-enriched testis. Endocrinology 101: 13–23

Tindall DJ, Rowley DR, Murthy L, Lipshultz LI, Chang CH (1985) Structure and biochemistry of the Sertoli cell. Int Rev Cytol 94: 127–149

Tsunenari I, Kast A (1992) Developmental and regressive changes in the testes of the Himalayan rabbit. Lab Anim 26: 167–179

Urbanski HF, Doan A, Pierce M, Fahrenbach WH, Collins PM (1992) Maturation of the hypothalamo-pituitary-gonadal axis of male Syrian hamsters. Biol Reprod 46: 991–996

van Haaster LH, de Rooij DG (1993) Spermatogenesis is accelerated in the immature Djungarian and Chinese hamster and rat. Biol Reprod 49: 1229–1235

Vergouwen RPFA, Jacobs SGPM, Huiskamp R, Davids JAG, de Rooij DG (1991) Proliferative activity of gonocytes, Sertoli cells and interstitial cells during testicular development in mice. J Reprod Fert 93: 233–243

Vergouwen RPFA, Huiskamp R, Bas RJ, Roepers-Gajadien HL, Davids JAG, de Rooij DG (1993) Postnatal development of testicular cell populations in mice. J Reprod Fert 99: 479–485

Viebahn C, Miething A (1991) Sertoli-Zell-Reifung: zeitlicher Zusammenhang zwischen morphologischen Veränderungen und biochemischem Umbau des Cytoskelettes im puberalen Kaninchenhoden. Anat Anz [Suppl] 170: 415–416

Vignon X, Terquem A, Dadoune JP (1987) The postnatal development of the junctional complexes of hamster Sertoli cells as revealed by HRP and freeze-fracture. J Submicrosc Cytol 19: 303–309

Vilar O, Perez del Cerro MI, Mancini RE (1962) The Sertoli cell as a "bridge cell" between the basal membrane and the germinal cells. Exp Cell Res 27: 158–161

Viswanathan N, Davis FC (1992) Timing of birth in Syrian hamsters. Biol Reprod 47: 6–10

Vitale R (1975) The development of the blood-testis barrier in Sertoli cell only rats. Anat Rec 181: 501

Vitale R, Fawcett DW, Dym M (1973) The normal development of the blood-testis barrier and the effects of clomiphene and estrogen treatment. Anat Rec 176: 333–344

Vogl AW, Pfeiffer DC, Redenbach DM, Grove BD (1993) Sertoli cell cytoskeleton. In: Russell LD, Griswold MD (eds) The Sertoli cell. Cache River, Clearwater, p 39–86

Vomachka AJ, Greenwald GS (1979) The development of gonadotropin and steroid hormone patterns in male and female hamsters from birth to puberty. Endocrinology 105: 960–966

von Ebner V (1871) Untersuchungen über den Bau der Samencanälchen und die Entwicklung der Spermatozoiden bei den Säugethieren und beim Menschen. In: Rollett A (ed) Untersuchungen aus dem Institute für Physiologie und Histologie in Graz, vol 2. Engelmann, Leipzig, p 200–236

von Ebner V (1888) Zur Spermatogenese bei den Säugethieren. Arch Mikr Anat 31: 236–292

Vornberger W, Prins G, Musto NA, SuarezQuian CA (1994) Androgen receptor distribution in rat testis: new implications for androgen regulation of spermatogenesis. Endocrinology 134: 2307–2316

Wartenberg H (1974) Spermatogenese – Oogenese: ein cyto-morphologischer Vergleich. Anat Anz [Suppl] 136: 63–92

Wartenberg H (1976) Comparative cytomorphologic aspects of the male germ cells, especially of the "gonia". Andrologia 8: 117–130

Wartenberg H (1990) Entwicklung der Genitalorgane und Bildung der Gameten. In: Hinrichsen KV (ed) Humanembryologie. Springer, Berlin Heidelberg New York, p 745–822

Wartenberg H, Schubert W (1976) Pinealorgan und Hoden: morphologische Untersuchungen zur Wirkung des Lichtentzuges auf die Spermatogenese des Goldhamsters. Anat Anz [Suppl] 140: 69–76

Wartenberg H, Hilscher B, Hilscher W (1998) Germ cell kinetics during early ovarian differentiation: an analysis of the oogonial cell cycle and the subsequent changes in oocyte development during the onset of meiosis in the rat. Microsc Res Tech (in press)

Weber JE, Russell LD (1987) A study of intercellular bridges during spermatogenesis in the rat. Am J Anat 180: 1–24

Weinbauer GF, Nieschlag E (1990) The role of testosterone in spermatogenesis. In: Nieschlag E, Behre HM (eds) Testosterone: action, deficiency, substitution. Springer, Berlin Heidelberg New York, p 23–50

Weissenberg R, Yossefi S, Oschry Y, Madgar I, Lewin LM (1994) Investigation of epididymal sperm maturation in the golden hamster. Int J Androl 17: 256–261

Welsh TH, Bambino TH, Hsueh AJW (1982) Mechanism of glucocorticoid-induced suppression of testicular androgen biosynthesis in vitro. Biol Reprod 27: 1138–1146

Wensing CJG (1986) Testicular descent in the rat and a comparison of this process in the rat with that in the pig. Anat Rec 214: 154–160

Whorwood CB, Franklyn JA, Sheppard MC, Stewart PM (1992) Tissue localization of 11β-hydroxysteroid dehydrogenase and its relationship to the glucocorticoid receptor. J Steroid Biochem Mol Biol 41: 21–28

Wong V, Russell LD (1983) Three-dimensional reconstruction of a rat stage V Sertoli cell: I. Methods, basic configuration, and dimensions. Am J Anat 167: 143–161

Wright WW, Frankel AI (1980) An androgen receptor in the nuclei of late spermatids in testes of male rats. Endocrinology 107: 314–318

Wrobel K-H, Bickel D, Kujat R, Schimmel M (1995) Configuration and distribution of bovine spermatogonia. Cell Tissue Res 279: 277–289

Wrobel K-H, Bickel D, Kujat R (1996) Immunohistochemical study of seminiferous epithelium in adult bovine testis using monoclonal antibodies against Ki-67 protein and proliferating cell nuclear antigen (PCNA). Cell Tissue Res 283: 191–201

Wyllie AH, Kerr JFR, Currie AR (1980) Cell death: the significance of apoptosis. Int Rev Cytol 68: 251–306

Ye SJ, Ying L, Ghosh S, de Franca LR, Russell LD (1993) Sertoli cell cycle: a re-examination of the structural changes during the cycle of the seminiferous epithelium of the rat. Anat Rec 237: 187–198

Zhengwei Y, Wreford NG, de Kretser DM (1990) A quantitative study of spermatogenesis in the developing rat testis. Biol Reprod 43: 629–635

Zhu L-J, Zong S-D, Phillips DM, Moo-Young AJ, Bardin CW (1997) Changes in the distribution of intermediate filaments in rat Sertoli cells during the seminiferous epithelium cycle and postnatal development. Anat Rec 248: 391–405

Zirkin BR, Santulli R, Awoniyi CA, Ewing LL (1989) Maintenance of advanced spermatogenic cells in the adult rat testis: quantitative relationship to testosterone concentration within the testis. Endocrinology 124: 3043–3049

Subject Index

Printing and binding: Druckerei Triltsch, Würzburg